Human Physiology

Human Physiology

Allyson Westley

Larsen & Keller
www.larsen-keller.com

Human Physiology
Allyson Westley
ISBN: 978-1-64172-114-1 (Hardback)

© 2019 Larsen & Keller

☰ Larsen & Keller

Published by Larsen and Keller Education,
5 Penn Plaza,
19th Floor,
New York, NY 10001, USA

Cataloging-in-Publication Data

Human physiology / Allyson Westley.
    p. cm.
Includes bibliographical references and index.
ISBN 978-1-64172-114-1
1. Human physiology. 2. Human anatomy. 3. Human biology. I. Westley, Allyson.
QP34.5 .H86 2019
612--dc23

This book contains information obtained from authentic and highly regarded sources. All chapters are published with permission under the Creative Commons Attribution Share Alike License or equivalent. A wide variety of references are listed. Permissions and sources are indicated; for detailed attributions, please refer to the permissions page. Reasonable efforts have been made to publish reliable data and information, but the authors, editors and publisher cannot assume any responsibility for the validity of all materials or the consequences of their use.

Trademark Notice: All trademarks used herein are the property of their respective owners. The use of any trademark in this text does not vest in the author or publisher any trademark ownership rights in such trademarks, nor does the use of such trademarks imply any affiliation with or endorsement of this book by such owners.

For more information regarding Larsen and Keller Education and its products, please visit the publisher's website www.larsen-keller.com

Table of Contents

Preface

Physiology is the branch of biology that studies normal mechanisms and their interactions within a living system. It focuses on how organisms, organ systems, cells and biomolecules interact and perform the chemical and physical functions in a living system. Human physiology explores the mechanical, physical and biochemical functions in humans. The human body is constituted by many interacting systems of organs. These systems include the nervous system, endocrine system, respiratory system, circulatory system, etc. The interactions of these systems are essential for the maintenance of homeostasis. Each system is important for the functioning of the body. This book is a compilation of chapters that discuss the most vital concepts in the field of human physiology. Different approaches and evaluations of human physiology have been included in this book. It is appropriate for those seeking detailed information in this area.

A detailed account of the significant topics covered in this book is provided below:

Chapter 1, Human physiology is the study of the functioning of the human body and includes the physical, mechanical, biochemical and bioelectrical functions of humans. This chapter will introduce briefly the significant aspects of the human physiology, such as cell physiology, blood physiology, digestive physiology, cardiovascular physiology, neurophysiology, etc. **Chapter 2**, The cell is the fundamental functional, structural and biological unit of living organisms. It consists of the cytoplasm, which is enclosed within a membrane, containing biomolecules such as nucleic acids and proteins. This chapter discusses in detail the various elements of cell structure and their functions. It includes topics such as cell nucleus, cytoplasm, cell junction, cell membrane and cell metabolism. **Chapter 3**, The endocrine system is an organ system in the human body that produces and secretes hormones for the regulation of the activity of cells and organs. This chapter has been carefully written to provide an easy understanding of the varied aspects of the endocrine system, the endocrine glands that make up the endocrine system and the mechanism of endocrine signaling. **Chapter 4**, The nervous system is an important system in a human body that controls the transmission of signals between different parts of the body. In vertebrates, it consists of two major systems, the peripheral and central nervous systems. The different aspects of the nervous system have been carefully analyzed in this chapter. **Chapter 5**, The muscular system of the human body is a major organ system that controls the movement of the body, circulates blood and maintains posture. This chapter closely examines the central constituents of the muscular system, such as smooth muscle, cardiac muscle, skeletal muscle, etc. **Chapter 6**, The human cardiovascular system consists of the heart, blood and blood vessels. Arteries, veins and capillaries are responsible for the circulation of blood throughout the human body. This chapter has been designed to provide an elaborate understanding of the cardiovascular system, circulation pathways, the heart and the cardiac cycle.

Chapter 7, The respiratory system is a biological system that consists of organs and structures for the exchange of gases in humans. The topics elaborated in this chapter on upper and lower respiratory system will help in providing a better perspective about the human respiratory system. **Chapter 8**, The urinary system of the human body consists of a pair of kidneys, ureters, the urinary bladder and the urethra. All these organs of the human urinary system have been extensively covered in this chapter. **Chapter 9**, The gastrointestinal system is an organ system, which is responsible for the digestion of food, extraction and absorption of energy and the expulsion of waste matter in the form of feces. This chapter delves into the major organs of the gastro intestinal system. **Chapter 10**, The reproductive system consists of sex organs that facilitate sexual reproduction in humans. This chapter provides an overview of the human reproductive system and includes valuable insights into the functions of the male and female reproductive system. **Chapter 11**, The immune system is a defense system that comprises of various biological structures and processes that enables an organism to fight against diseases and infections. This chapter elaborates the relevant aspects of the innate and adaptive immune system which will help in developing a holistic understanding of the human immune system.

It gives me an immense pleasure to thank our entire team for their efforts. Finally in the end, I would like to thank my family and colleagues who have been a great source of inspiration and support.

Allyson Westley

Introduction to Human Physiology

Human physiology is the study of the functioning of the human body and includes the physical, mechanical, biochemical and bioelectrical functions of humans. This chapter will introduce briefly the significant aspects of the human physiology, such as cell physiology, blood physiology, digestive physiology, cardiovascular physiology, neurophysiology, etc.

Physiology

Physiology is the branch of biology relating to the function of organs and organ systems, and how they work within the body to respond to challenges. It covers life from the single cell, where it overlaps with biochemistry and molecular biology, through questions about how individual organs work (e.g. heart, lungs, kidneys) right up to the whole-organism level, where physiologists tackle questions about hormonal influences on behaviour and the function of the brain. Physiology, therefore, has something to say about every aspect of life: our integrated approach makes physiologists invaluable contributors in studies ranging from genetics to psychology.

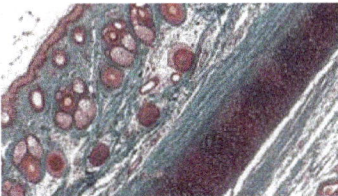

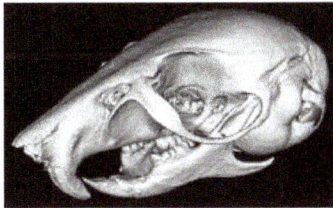

In its applied aspects, physiology deals with the function and malfunction of parts of the human body with reference to health and disease (areas relating to medicine), how to improve crop yield (areas relating to plant sciences) as well as the practical problems of animal, plant and microbial performance and their responses to challenging conditions (areas relating to ecology).

In science at the moment, there is a tendency to look downwards rather than upwards, at molecular mechanisms in preference to the often less tractable problems posed by systems as a whole. Physiology is no exception, and the temptations to concentrate on molecular aspects of the subject - areas where new information is easier to come by, and where conceptual problems are less obvious - have never been stronger. But, as

Research Councils are keen to emphasize, the largest gaps in our knowledge are often how the molecules translate into the function - and malfunction - of the organism as a whole. These questions are difficult to answer and sometimes, as in the case of the brain, difficult to formulate as well. Part of the training of a physiologist is to learn to think, argue and to see problems on a wider scale, without losing sight of the whole organism.

Here are some key points about physiology:

- Physiology can be considered a study of the functions and processes that create life.

- The study of physiology can be traced back to at least 420 BC.

- The study of physiology is split into many disciplines covering topics as different as exercise, evolution, and defense.

Human Physiology

Human physiology is a life science and a branch of animal physiology. It is specifically the study of how systems of the body function in a well state, and this analysis of function is often at the cellular level, not of single cells but of how cells work in concert to achieve a normal state of function. Basic human physiology studies the body's systems that function appropriately and as expected, while other disciplines like pathophysiology may look at the way body systems develop disease in attempts to find insight into how to cure diseases.

There may be several main concerns in human physiology from a scientific standpoint. These concerns include the way interdependence between body systems occurs (such as the central nervous system and the musculoskeletal system). This is called integration.

Another point of interest is communication, which is how the body's systems send signals to function in specific ways. These signals could be electrical impulses or the release of chemicals. Lastly, the physiologist wants to define and observe homeostasis, in any of the systems studied. In other words, how does the body maintain a normal state, and what are the processes by which it does so?

It might be oversimplification to say that human physiology attempts to answer the question of "how things work." However, this is fairly accurate, and it's an important question to answer. Understanding the normal function of the body's systems is valuable because it establishes baselines for understanding what is abnormal. It is very difficult to diagnose disease unless a clear deviation from the norm can be determined, and therefore establishing this norm is of great value in medicine and in human health.

For instance, over time, physiology and biochemistry have helped to determine what constitutes normal blood levels of certain substances. When something like sugar levels become too high, it may have impact on various systems in the body and be indication of diseases like diabetes. Only by knowing baseline levels for various sugar types in blood, doctors can determine whether diabetes is present. This knowledge has been extrapolated to allow patients to keep records of their own blood sugar at home. With testing they can be assured that they are regulating blood sugar appropriately or they can make medication adjustments when blood sugar levels are too high or too low.

It's suggested that early studies in human physiology and anatomy began over 2000 years ago, and names like Hippocrates and Aristotle are usually given as early physiologists. The trouble with early thought was it didn't allow for many examinations of humans, and most humans examined were dead. The idea of cells wouldn't be posited until much later in history. Much more was done in the field of anatomy, which is an intricately related discipline to human physiology, that describes the forms present in the body, and yet again, unless these forms were obvious and on the surface, they typically didn't get much exploration unless a person was dead.

More studies were possible on animals, and actually animal physiology is still used and extrapolated to human beings all of the time. Even today when medical science is much more delicate, most well humans would not consent to studies of some of the ways their body systems work. Few people would volunteer to have abnormal rhythms of their heart induced as part of electrophysiological cardiology studies to determine what causes arrhythmias, as this might be dangerous. However electrophysiologists can induce arrhythmias in animals to determine what factors destroy balance in the electrical system in the heart.

Over time, human physiology has helped to define the major systems of the body and how they work to achieve wellness. Basic introductory courses tend to look at each of these systems, which may be roughly defined as the following: circulatory, respiratory, endocrine, reproductive, immune, musculoskeletal, nervous, integumentary, renal, and gastrointestinal.

While breaking the body into systems can help describe function, it isn't always so neat from a scientific standpoint. Systems are interdependent on each other. Lose renal or respiratory function, and everything else becomes affected. Moreover, many vital organs or parts of the body may participate in several systems.

Biological Systems

The major systems covered in the study of human physiology are as follows:

- Circulatory system: Including the heart, the blood vessels, properties of the blood, and how circulation works in sickness and health.

- Digestive/excretory system: Charting the movement of solids from the mouth to the anus; this includes study of the spleen, liver, and pancreas, the conversion of food into fuel and its final exit from the body.

- Endocrine system: The study of endocrine hormones that carry signals throughout the organism, helping it to respond in concert. The principal endocrine glands - the pituitary, thyroid, adrenals, pancreas, parathyroids, and gonads - are a major focus, but nearly all organs release endocrine hormones.

- Immune system: The body's natural defense system is comprised of white blood cells, the thymus, and lymph systems. A complex array of receptors and molecules combine to protect the host from attacks by pathogens. Molecules such as antibodies and cytokines feature heavily.

- Integumentary system: The skin, hair, nails, sweat glands, and sebaceous glands (secreting an oily or waxy substance).

- Musculoskeletal system: The skeleton and muscles, tendons, ligaments, and cartilage. Bone marrow - where red blood cells are made - and how bones store calcium and phosphate are included.

- Nervous system: The central nervous system (brain and spinal cord) and the peripheral nervous system. Study of the nervous system includes research into the senses, memory, emotion, movement, and thought.

- Renal/urinary system: Including the kidneys, ureters, bladder, and urethra, this system removes water from the blood, produces urine, and carries away waste.

- Reproductive system: Consisting of the gonads and the sex organs. Study of this system also includes investigating the way a fetus is created and nurtured for 9 months.

- Respiratory system: Consisting of the nose, nasopharynx, trachea, and lungs. This system brings in oxygen and expels carbon dioxide and water.

Branches

Defense physiology investigates nature's natural defensive reactions.

There are a great number of disciplines that use the word physiology in their title. Below are some examples:

- Cell physiology: Studying the way cells work and interact; cell physiology mostly concentrates on membrane transport and neuron transmission.

- Systems physiology: This focuses on the computational and mathematical modeling of complex biological systems. It tries to describe the way individual cells or components of a system converge to respond as a whole. They often investigate metabolic networks and cell signaling.

- Evolutionary physiology: Studying the way systems, or parts of systems, have adapted and changed over multiple generations. Research topics cover a lot of ground including the role of behavior in evolution, sexual selection, and physiological changes in relation to geographic variation.

- Defense physiology: Changes that occur as a reaction to a potential threat, such as preparation for the fight-or-flight response.

- Exercise physiology: As the name suggests, this is the study of the physiology of physical exercise. This includes research into bioenergetics, biochemistry, cardiopulmonary function, biomechanics, hematology, skeletal muscle physiology, neuroendocrine function, and nervous system function.

The topics mentioned above are just a small selection of the available physiologies. The field of physiology is as essential as it is vast.

Functions of Human Life

The different organ systems each have different functions and therefore unique roles to perform in physiology. These many functions can be summarized in terms of a few that we might consider definitive of human life: organization, metabolism, responsiveness, movement, development, and reproduction.

Organization

A human body consists of trillions of cells organized in a way that maintains distinct internal compartments. These compartments keep body cells separated from external environmental threats and keep the cells moist and nourished. They also separate internal body fluids from the countless microorganisms that grow on body surfaces, including the lining of certain tracts, or passageways. The intestinal tract, for example, is home to even more bacteria cells than the total of all human cells in the body, yet these bacteria are outside the body and cannot be allowed to circulate freely inside the body.

Cells, for example, have a cell membrane (also referred to as the plasma membrane) that keeps the intracellular environment-the fluids and organelles-separate from the extracellular environment. Blood vessels keep blood inside a closed circulatory system, and nerves and muscles are wrapped in connective tissue sheaths that separate them from surrounding structures. In the chest and abdomen, a variety of internal membranes keep major organs such as the lungs, heart, and kidneys separate from others.

The body's largest organ system is the integumentary system, which includes the skin and its associated structures, such as hair and nails. The surface tissue of skin is a barrier that protects internal structures and fluids from potentially harmful microorganisms and other toxins.

Metabolism

The first law of thermodynamics holds that energy can neither be created nor destroyed—it can only change form. Your basic function as an organism is to consume (ingest) energy and molecules in the foods you eat, convert some of it into fuel for movement, sustain your body functions, and build and maintain your body structures. There are two types of reactions that accomplish this: anabolism and catabolism.

- Anabolism is the process whereby smaller, simpler molecules are combined into larger, more complex substances. Your body can assemble, by utilizing energy, the complex chemicals it needs by combining small molecules derived from the foods you eat.

- Catabolism is the process by which larger more complex substances are broken down into smaller simpler molecules. Catabolism releases energy. The complex molecules found in foods are broken down so the body can use their parts to assemble the structures and substances needed for life.

Taken together, these two processes are called metabolism. Metabolism is the sum of all anabolic and catabolic reactions that take place in the body. Both anabolism and catabolism occur simultaneously and continuously to keep you alive.

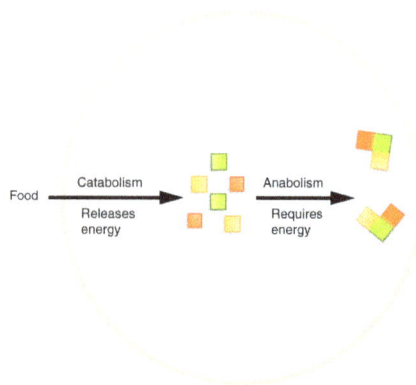

Figure: Metabolism. Anabolic reactions are building reactions, and they consume energy.
Catabolic reactions break materials down and release energy.
Metabolism includes both anabolic and catabolic reactions.

Every cell in your body makes use of a chemical compound, adenosine triphosphate (ATP), to store and release energy. The cell stores energy in the synthesis (anabolism) of ATP, then moves the ATP molecules to the location where energy is needed to fuel cellular activities. Then, the ATP is broken down (catabolism) and a controlled amount of energy is released, which is used by the cell to perform a particular job.

Responsiveness

Responsiveness is the ability of an organism to adjust to changes in its internal and external environments. An example of responsiveness to external stimuli could include moving toward sources of food and water and away from perceived dangers. Changes in an organism's internal environment, such as increased body temperature, can cause the responses of sweating and the dilation of blood vessels in the skin in order to decrease body temperature.

Movement

Human movement includes not only actions at the joints of the body, but also the motion of individual organs and even individual cells. Red and white blood cells are moving throughout your body, muscle cells are contracting and relaxing to maintain your posture and to focus your vision, and glands are secreting chemicals to regulate body functions. Your body is coordinating the action of entire muscle groups to enable you to move air into and out of your lungs, to push blood throughout your body, and to propel the food you have eaten through your digestive tract. Consciously, of course, you contract your skeletal muscles to move the bones of your skeleton to get from one place to another (as the runners are doing in figure), and to carry out all of the activities of your daily life.

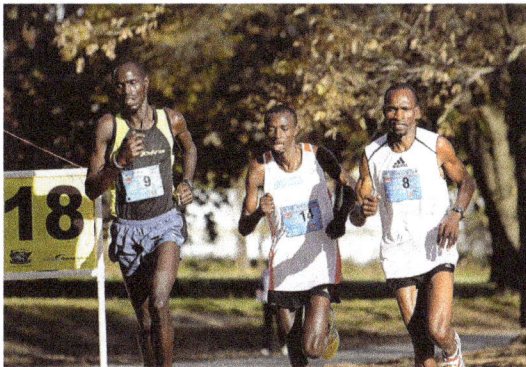

Figure: Marathon Runners

Runners demonstrate two characteristics of living humans—responsiveness and movement. Anatomic structures and physiological processes allow runners to coordinate the action of muscle groups and sweat in response to rising internal body temperature.

Development, Growth and Reproduction

Development is all of the changes the body goes through in life. Development includes the process of differentiation, in which unspecialized cells become specialized in structure and function to perform certain tasks in the body. Development also includes the processes of growth and repair, both of which involve cell differentiation.

Growth is the increase in body size. Humans, like all multicellular organisms, grow by increasing the number of existing cells, increasing the amount of non-cellular material around cells (such as mineral deposits in bone), and, within very narrow limits, increasing the size of existing cells.

Reproduction is the formation of a new organism from parent organisms. In humans, reproduction is carried out by the male and female reproductive systems. Because death will come to all complex organisms, without reproduction, the line of organisms would end.

Cell Physiology

Cell physiology is a field of biology which focuses on studying the function of cells, and how cells interact with each other and with the larger organism they inhabit. Researchers in this field use a variety of tools to observe cells and learn more about their structure and function, such as microscopes and more advanced imaging equipment to see the structures inside of cells in greater detail.

Each cell within an organism is designed to act as an independently functioning unit which supports the larger organism as a whole. Cell physiology looks at the structure of different types of cells, and how cells function. It also looks at how cells come together to create organs and other structures, and how the cells within an organism work together. All of the normal functions of a cell are covered in this field, as are the many different types of cells which can be found in a single organism.

Understanding cell physiology is important to the understanding of larger organisms, as many important activities take place at the cellular level. By studying how cells are supposed to work under normal conditions, researchers can also make it easy to identify errors, problems, and malfunctions when the cells in an organism become abnormal. Identifying abnormalities and their causes can be valuable in the treatment and management of disease, and for the general advancement of biology as a science.

Blood Physiology

Blood is essential to life. Blood circulates through our body and delivers essential substances like oxygen and nutrients to the body's cells. It also transports metabolic waste products away from those same cells. There is no substitute for blood. It cannot be made or manufactured. Generous blood donors are the only source of blood for patients in need of a blood transfusion.

Blood Components

There are four basic components that comprise human blood: plasma, red blood cells, white blood cells and platelets.

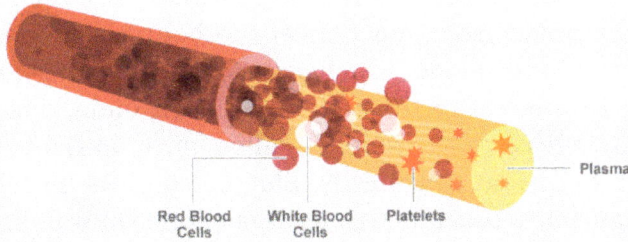

Red Blood Cells   White Blood Cells   Platelets   Plasma

Red Blood Cells

Red blood cells represent 40%-45% of the blood volume. They are generated from the bone marrow at a rate of four to five billion per hour. They have a lifecycle of about 120 days in the body.

Platelets

Platelets are an amazing part of the blood. Platelets are the smallest of our blood cells and literally look like small plates in their non-active form. Platelets control bleeding. Wherever a wound occurs, the blood vessel will send out a signal. Platelets receive that signal and travel to the area and transform into their "active" formation, growing long tentacles to make contact with the vessel and form clusters to plug the wound until it heals.

Plasma

Plasma is the liquid portion of the blood. Plasma is yellowish in color and is made up mostly of water, but it also contains proteins, sugars, hormones and salts. It transports water and nutrients to your body's tissues.

White Blood Cells

Although white blood cells (leukocytes) only account for about 1% of the blood, they are very important. White blood cells are essential for good health and protection against illness and disease. Like red blood cells, they are constantly being generated from your bone marrow. They flow through the bloodstream and attack foreign bodies, like viruses and bacteria. They can even leave the bloodstream to extend the fight into tissue.

Functions of Blood

The primary function of blood is to deliver oxygen and nutrients to and remove wastes from body cells, but that is only the beginning of the story. The specific functions of blood also include defense, distribution of heat, and maintenance of homeostasis.

Transportation

Nutrients from the foods you eat are absorbed in the digestive tract. Most of these travel in the bloodstream directly to the liver, where they are processed and released back into the bloodstream for delivery to body cells. Oxygen from the air you breathe diffuses into the blood, which moves from the lungs to the heart, which then pumps it out to the rest of the body. Moreover, endocrine glands scattered throughout the body release their products, called hormones, into the bloodstream, which carries them to distant target cells. Blood also picks up cellular wastes and byproducts, and transports them to various organs for removal. For instance, blood moves carbon dioxide to the lungs for exhalation from the body, and various waste products are transported to the kidneys and liver for excretion from the body in the form of urine or bile.

Defense

Many types of WBCs protect the body from external threats, such as disease-causing bacteria that have entered the bloodstream in a wound. Other WBCs seek out and destroy internal threats, such as cells with mutated DNA that could multiply to become cancerous, or body cells infected with viruses.

When damage to the vessels results in bleeding, blood platelets and certain proteins dissolved in the plasma, the fluid portion of the blood, interact to block the ruptured areas of the blood vessels involved. This protects the body from further blood loss.

Maintenance of Homeostasis

Recall that body temperature is regulated via a classic negative-feedback loop. If you were exercising on a warm day, your rising core body temperature would trigger several homeostatic mechanisms, including increased transport of blood from your core to your body periphery, which is typically cooler. As blood passes through the vessels of the skin, heat would be dissipated to the environment, and the blood returning to your body core would be cooler. In contrast, on a cold day, blood is diverted away from the skin to maintain a warmer body core. In extreme cases, this may result in frostbite.

Blood also helps to maintain the chemical balance of the body. Proteins and other compounds in blood act as buffers, which thereby help to regulate the pH of body tissues. Blood also helps to regulate the water content of body cells.

Composition of Blood

You have probably had blood drawn from a superficial vein in your arm, which was then sent to a lab for analysis. Some of the most common blood tests—for instance, those measuring lipid or glucose levels in plasma—determine which substances are present within blood and in what quantities. Other blood tests check for the composition of the blood itself, including the quantities and types of formed elements.

One such test, called a hematocrit, measures the percentage of RBCs, clinically known as erythrocytes, in a blood sample. It is performed by spinning the blood sample in a specialized centrifuge, a process that causes the heavier elements suspended within the blood sample to separate from the lightweight, liquid plasma. Because the heaviest elements in blood are the erythrocytes, these settle at the very bottom of the hematocrit tube. Located above the erythrocytes is a pale, thin layer composed of the remaining formed elements of blood. These are the WBCs, clinically known as leukocytes, and the platelets, cell fragments also called thrombocytes. This layer is referred to as the buffy coat because of its color; it normally constitutes less than 1 percent of a blood sample. Above the buffy coat is the blood plasma, normally a pale, straw-colored fluid, which constitutes the remainder of the sample.

The volume of erythrocytes after centrifugation is also commonly referred to as packed cell volume (PCV). In normal blood, about 45 percent of a sample is erythrocytes. The hematocrit of any one sample can vary significantly, however, about 36–50 percent, according to gender and other factors. Normal hematocrit values for females range from 37 to 47, with a mean value of 41; for males, hematocrit ranges from 42 to 52, with a mean of 47. The percentage of other formed elements, the WBCs and platelets, is extremely small so it is not normally considered with the hematocrit. So, the mean plasma percentage is the percent of blood that is not erythrocytes: for females, it is approximately 59 (or 100 minus 41), and for males, it is approximately 53 (or 100 minus 47).

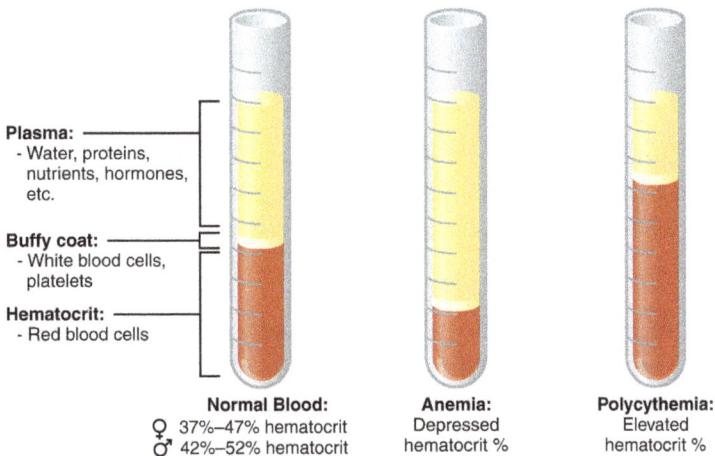

Plasma:
- Water, proteins, nutrients, hormones, etc.

Buffy coat:
- White blood cells, platelets

Hematocrit:
- Red blood cells

Normal Blood:
♀ 37%–47% hematocrit
♂ 42%–52% hematocrit

Anemia:
Depressed
hematocrit %

Polycythemia:
Elevated
hematocrit %

Figure: composition of blood

The cellular elements of blood include a vast number of erythrocytes and comparatively fewer leukocytes and platelets. Plasma is the fluid in which the formed elements are suspended. A sample of blood spun in a centrifuge reveals that plasma is the lightest component. It floats at the top of the tube separated from the heaviest elements, the erythrocytes, by a buffy coat of leukocytes and platelets. Hematocrit is the percentage of the total sample that is comprised of erythrocytes. Depressed and elevated hematocrit levels are shown for comparison.

Characteristics of Blood

When you think about blood, the first characteristic that probably comes to mind is its color. Blood that has just taken up oxygen in the lungs is bright red, and blood that has released oxygen in the tissues is a more dusky red. This is because hemoglobin is a pigment that changes color, depending upon the degree of oxygen saturation.

Blood is viscous and somewhat sticky to the touch. It has a viscosity approximately five times greater than water. Viscosity is a measure of a fluid's thickness or resistance to flow, and is influenced by the presence of the plasma proteins and formed elements within the blood. The viscosity of blood has a dramatic impact on blood pressure and flow. Consider the difference in flow between water and honey. The more viscous honey would demonstrate a greater resistance to flow than the less viscous water. The same principle applies to blood.

The normal temperature of blood is slightly higher than normal body temperature—about 38°C (or 100.4°F), compared to 37°C (or 98.6°F) for an internal body temperature reading, although daily variations of 0.5°C are normal. Although the surface of blood vessels is relatively smooth, as blood flows through them, it experiences some friction and resistance, especially as vessels age and lose their elasticity, thereby producing heat. This accounts for its slightly higher temperature.

The pH of blood averages about 7.4; however, it can range from 7.35 to 7.45 in a healthy person. Blood is therefore, somewhat more basic (alkaline) on a chemical scale than pure water, which has a pH of 7.0. Blood contains numerous buffers that actually help to regulate pH.

Blood constitutes approximately 8 percent of adult body weight. Adult males typically average about 5 to 6 liters of blood. Females average 4–5 liters.

Blood Plasma

Like other fluids in the body, plasma is composed primarily of water: In fact, it is about 92 percent water. Dissolved or suspended within this water is a mixture of substances, most of which are proteins. There are literally hundreds of substances dissolved or suspended in the plasma, although many of them are found only in very small quantities.

Plasma Proteins

About 7 percent of the volume of plasma—nearly all that is not water—is made of proteins. These include several plasma proteins (proteins that are unique to the plasma), plus a much smaller number of regulatory proteins, including enzymes and some hormones. The major components of plasma are summarized in figure.

The three major groups of plasma proteins are as follows:

- Albumin is the most abundant of the plasma proteins. Manufactured by the liver, albumin molecules serve as binding proteins—transport vehicles for fatty acids and steroid hormones. Recall that lipids are hydrophobic; however, their binding to albumin enables their transport in the watery plasma. Albumin is also the most significant contributor to the osmotic pressure of blood; that is, its presence holds water inside the blood vessels and draws water from the tissues, across blood vessel walls, and into the bloodstream. This in turn helps to maintain both blood volume and blood pressure. Albumin normally accounts for approximately 54 percent of the total plasma protein content, in clinical levels of 3.5–5.0 g/dL blood.

- The second most common plasma proteins are the globulins. A heterogeneous group, there are three main subgroups known as alpha, beta, and gamma globulins. The alpha and beta globulins transport iron, lipids, and the fat-soluble vitamins A, D, E, and K to the cells; like albumin, they also contribute to osmotic pressure. The gamma globulins are proteins involved in immunity and are better known as an antibodies or immunoglobulins. Although other plasma proteins are produced by the liver, immunoglobulins are produced by specialized leukocytes known as plasma cells. Globulins make up approximately 38 percent of the total plasma protein volume, in clinical levels of 1.0–1.5 g/dL blood.

- The least abundant plasma protein is fibrinogen. Like albumin and the alpha and beta globulins, fibrinogen is produced by the liver. It is essential for blood clotting. Fibrinogen accounts for about 7 percent of the total plasma protein volume, in clinical levels of 0.2–0.45 g/dL blood.

Other Plasma Solutes

In addition to proteins, plasma contains a wide variety of other substances. These include various electrolytes, such as sodium, potassium, and calcium ions; dissolved gases, such as oxygen, carbon dioxide, and nitrogen; various organic nutrients, such as vitamins, lipids, glucose, and amino acids; and metabolic wastes. All of these non-protein solutes combined contribute approximately 1 percent to the total volume of plasma.

Component and % of blood	Subcomponent and % of component	Type and % (where appropriate)	Site of production	Major function(s)
Plasma 46–63 percent	Water 92 percent	Fluid	Absorbed by intestinal tract or produced by metabolism	Transport medium
	Plasma proteins 7 percent	Albumin 54–60 percent	Liver	Maintain osmotic concentration, transport lipid molecules
		Globulins 35–38 percent	Alpha globulins— liver	Transport, maintain osmotic concentration
			Beta globulins— liver	Transport, maintain osmotic concentration
			Gamma globulins (immunoglobulins) —plasma cells	Immune responses
		Fibrinogen 4–7 percent	Liver	Blood clotting in hemostasis
	Regulatory proteins <1 percent	Hormones and enzymes	Various sources	Regulate various body functions
	Other solutes 1 percent	Nutrients, gases, and wastes	Absorbed by intestinal tract, exchanged in respiratory system, or produced by cells	Numerous and varied

Formed elements 37–54 percent	Erythrocytes 99 percent	Erythrocytes	Red bone marrow	Transport gases, primarily oxygen and some carbon dioxide
	Leukocytes <1 percent Platelets <1 percent	Granular leukocytes: neutrophils eosinophils basophils	Red bone marrow	Nonspecific immunity
		Agranular leukocytes: lymphocytes monocytes	Lymphocytes: bone marrow and lymphatic tissue	Lymphocytes: specific immunity
			Monocytes: red bone marrow	Monocytes: nonspecific immunity
	Platelets <1 percent		Megakaryocytes: red bone marrow	Hemostasis

Major blood components

Phlebotomy and Medical Lab Technology: Phlebotomists are professionals trained to draw blood (phleb- = "a blood vessel"; -tomy = "to cut"). When more than a few drops of blood are required, phlebotomists perform a venipuncture, typically of a surface vein in the arm. They perform a capillary stick on a finger, an earlobe, or the heel of an infant when only a small quantity of blood is required. An arterial stick is collected from an artery and used to analyze blood gases. After collection, the blood may be analyzed by medical laboratories or perhaps used for transfusions, donations, or research. While many allied health professionals practice phlebotomy, the American Society of Phlebotomy Technicians issues certificates to individuals passing a national examination, and some large labs and hospitals hire individuals expressly for their skill in phlebotomy.

Medical or clinical laboratories employ a variety of individuals in technical positions:

- Medical technologists (MT), also known as clinical laboratory technologists (CLT), typically hold a bachelor's degree and certification from an accredited training program. They perform a wide variety of tests on various body fluids, including blood. The information they provide is essential to the primary care providers in determining a diagnosis and in monitoring the course of a disease and response to treatment.

- Medical laboratory technicians (MLT) typically have an associate's degree but may perform duties similar to those of an MT.

- Medical laboratory assistants (MLA) spend the majority of their time processing samples and carrying out routine assignments within the lab. Clinical training is required, but a degree may not be essential to obtaining a position.

Digestive Physiology

Normal digestive physiology is essential for breaking down food into its basic components so that nutrients can enter the blood stream. Digestion begins by chewing food, which starts to get broken down by saliva. Food then goes down the esophagus into the stomach, and then to the large intestine. The liver and pancreas are the only solid organs in the digestive system, and both supply fluids to the intestine to further digest food. Sugars, amino acids, and fatty acids are absorbed through the lining of the small intestine into the blood, while all muscle contractions in the digestive system are controlled by nerves both outside and within each organ.

Digestion begins as soon as food enters the mouth. The salivary glands secrete fluid filled with enzymes, mucus, electrolytes, and water. Potassium and bicarbonates are released in the saliva ducts, which help to regulate the acid produced in the stomach. Chewing breaks down and softens food to make it easier for enzymes to work. Swallowing is aided by the tongue and the peristaltic contractions of the esophagus, which are controlled by muscle structures called sphincters.

The stomach is the component of digestive physiology where food is liquefied. It can contract and expand depending on consumption, while enzymes and muscle contractions aid digestion. The lower part of the stomach allows liquefied food to pass into the small intestine where fluids from the pancreas and liver mix in, such as bile, which serves to dissolve fat. Fluids from the pancreas break down fat, protein, and carbohydrates.

What is left in the small intestine is either absorbed into the blood or passes through as waste. Just about all nutrients are passed into the blood here, including electrolytes such as sodium, potassium, chloride, and organic molecules. What is absorbed by the small intestine travels through the blood to the liver via the portal vein, where vitamins are stored and the release of glucose into the blood is controlled. The liver also metabolizes fat and protein and is responsible for the storage and distribution of fat, so it is a crucial part of digestive physiology.

In the large intestine, any water and electrolytes that are left are absorbed. What remains in this part of digestive physiology is dehydrated, while bacteria and mucus are mixed in to form feces. Microbial organisms break down cellulose and carbohydrates, and any fatty acids and vitamin K left are absorbed and utilized for metabolism.

Cardiovascular Physiology

Cardiovascular physiology is the study of the heart and circulatory system. A number of medical professionals rely on cardiovascular physiology in their work, including pulmonologists, cardiologists, and cardiothoracic surgeons. Because the cardiovascular system is so important to healthy function of an organism, it is also covered extensively during medical education, ensuring that all medical practitioners understand the basic anatomy and physiology of the circulatory system and lungs.

A number of very complex processes are involved in the function of the heart and circulatory system, from the osmosis which allows blood to deliver fresh oxygen and nutrients to the body while picking up waste materials for elimination to the electrical impulses which keep the heart beating. Cardiovascular physiology includes an extensive examination of all of these processes, including the chemistry of the blood, the physical anatomy of the heart and vascular system, and the role of the lungs in the oxygenation of blood.

Abnormalities in physiology are also of interest, ranging from diseases which alter blood pressure to congenital heart abnormalities which interfere with normal heart function. Cardiovascular pathology, the study of such abnormalities, also includes an understanding of techniques which can be used to address or monitor them, including the introduction of medications to regulate blood pressure, and artificial pacemakers to normalize heart rhythm.

A knowledge of cardiovascular physiology can allow a doctor to make a treatment recommendation to a patient, and it can also be used for things like patient education, public health initiatives which are designed to address rising rates of heart disease, and medical imaging studies which assess heart and circulatory function. The vascular system is also of interest to pharmacologists, as it can be used as a highly effective delivery system for medications. Cardiovascular physiology is also a topic of concern to some body workers, who can influence cardiovascular function with their work, as seen when massage therapy inadvertently causes fluid retention by interfering with circulation.

Respiration

An understanding of the physiology of respiration will enable an increased understanding of some of the disease processes encountered. It is worth spending some time trying to understand some of the concepts involved.

The main purpose behind respiration is to supply the cells with oxygen and to remove carbon dioxide, the waste product. There are three key elements to this process: pulmonary ventilation, external (pulmonary) respiration, and internal (tissue) respiration.

Pulmonary Ventilation

Inspiration is the process by which we breathe in. The pressure inside the lungs before each breath equals about 760 mm Hg or 1 atmosphere at sea level.

The pressure inside the lungs needs to be lower than this for air to enter and this is achieved by increasing the volume of the thoracic cavity. Boyles law states that the pressure of a gas in a closed container is inversely proportional to the volume of the container.

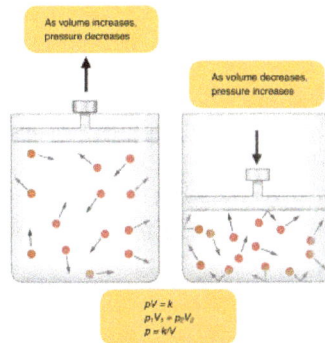

As volume increases, pressure decreases

As volume decreases, pressure increases

$$pV = k$$
$$p_1V_1 = p_2V_2$$
$$p = k/V$$

So, the difference in pressure created when we breathe in forces air into our lungs.

In order to increase the volume the principal muscles must contract. These muscles are the diaphragm and the intercostals or those muscles between the ribs.

The diaphragm is a dome shaped skeletal muscle which is innervated by the phrenic nerve, which emerges from spinal cord at cervical levels 3, 4, and 5. Contraction of the diaphragm causes it to flatten which increases the vertical dimension of the thoracic cavity and therefore its volume.

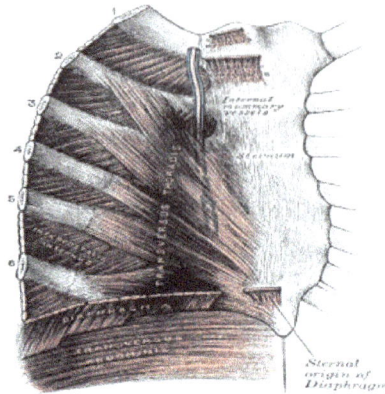

The diaphragm moves from 1cm to 10cm depending on whether there is normal or heavy breathing. Obesity and pregnancy can both obstruct this process leading to breathlessness.

As the diaphragm is being pulled down the external intercostals also contract pulling the ribs upwards.

The pressure between the two pleural membranes is also important to consider. During normal breathing this intrapleural pressure is always below atmospheric pressure.

As the diaphragm contracts the intrapleural pressure falls from 756 mm Hg to 754 mm Hg and this pressure change pulls the walls of the lungs outwards.

The two pleura adhere strongly to each other because of the below atmospheric pressure and the surface tension between them.

So the combination of the diaphragm and the external intercostals help to increase the volume of the thoracic cavity, which then drops the pressure within the cavity.

The alveolar pressure drops from 760 mm Hg to 758 mm Hg, and the pressure gradient created between the inside of the lungs and the atmosphere draws air in.

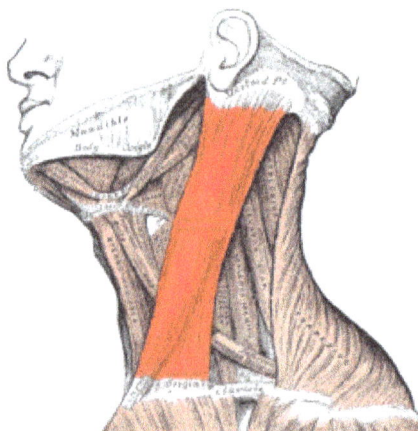

During deep breathing, caused by exertion or lung disease other muscles become involved in inspiration and these are known as the accessory muscles. These include the sternomastoid, scalenes and pectoralis major.

Expiration is the reversal of the pressure gradient. This is normally a passive process depending on relaxation of the muscles involved in inspiration. It does however depend upon the elastic recoil of the lung and the inward pull of the surface tension due to the alveolar fluid.

As the external intercostals and the diaphragm relax, the ribs move downwards and the dome of the diaphragm moves back up. This reduces the volume of the thoracic cavity and hence the pressure increases to around 762 mm Hg.

The surface tension in the alveoli also creates a pull which causes the bronchioles and the alveolar ducts to recoil.

Air then flows out of the lung.

During active expiration, when breathing becomes laboured, the abdominal and internal intercostal muscles become involved in further flattening the dome of the diaphragm to help expel the air.

Physiology of External Respiration

Deoxygenated blood enters the lungs from the right side of the heart. Oxygenated blood then leave the lungs to enter the left side of the heart. Atmospheric air enters the lungs and moves into the alveoli. The partial pressure of oxygen in deoxygenated blood is only 40 mmHg, whilst the partial pressure of oxygen in atmospheric air in the alveoli is 105 mmHg. Because of the pressure gradient created between these different partial pressures of oxygen will move from an area of high concentration to an area of low concentration. In this case moving from the alveoli across the membrane and into the pulmonary capillaries.

The partial pressure of carbon dioxide in alveoli air is 40 mmHg whilst in the deoxygenated blood the partial pressure of carbon dioxide is 45 mmHg. Again, because of the pressure gradient created, carbon dioxide will therefore move from an area of high concentration to an area of low concentration. So the carbon dioxide moves from the pulmonary capillary into the alveoli.

The rate of respiration will depend on several factors:

1. The rate of diffusion across any membrane is directly proportional to the size of that membrane, or its surface area. The surface area of the lung is approximately 70 m^2, which provides a large area over which the gas can exchange. This is surface area is reduced then gas exchange becomes more difficult. An example of a condition which may cause this would be emphysema where the alveoli walls disintegrate and the surface area of the lung is therefore reduced.

2. The thickness of the membrane is also crucial. The thick of the membrane the harder it is for the gas to diffuse across. This is why the total thickness of the alveoli are-capillaries membranes is only about 0.5 micro metres. Anything which increases this distance will also make gas exchange more difficult. So for example a buildup of fluid or lung secretions as in a severe pneumonia will worsen the gas exchange.

3. The partial pressure difference of the gases will also affect how well respiration can take place. If the partial pressure is reduced, when for example one goes to a higher altitude, then the rate of gas exchange will slow down giving the typical symptoms of high altitude sickness caused by the low partial pressure of oxygen in the blood.

Physiology of Internal Respiration

The process of gas exchange between the tissue blood capillaries and tissue cells is called internal respiration. This also occurs because of the partial pressure differences between the gases on either side of the membrane. Oxygenated blood in the tissue capillaries has a partial pressure of 105 mmHg where is that in the tissue as a partial pressure of oxygen of only 40 mmHg. So once again the gas moves from an area of high pressure to an area of low pressure. The same occurs with carbon dioxide at the tissue level, with this obviously moving in opposition to the oxygen.

Neural Control of Respiration

Although gas exchange takes place in the lungs, the respiratory system is controlled by the central nervous system (CNS). While we do have some voluntary control of breathing, it is regulated automatically and functions whether we think about it or not. Breathing can, however, be suppressed at the neurological level due to narcotic or sedative overdose, as well as brainstem injury.

The portions of the CNS that control respiration are located within the brain stem—specifically within the pons and the medulla. These components are responsible for the nerve impulses, which are transmitted via the phrenic and other motor nerves to the diaphragm and intercostal muscles, controlling our basic breathing rhythm. Also located in the brainstem are the central chemoreceptors. These specialized cells signal the body to adjust ventilation based indirectly on arterial CO_2 ($PaCO_2$) levels. This accounts for our primary respiratory drive. The peripheral chemoreceptors, which are located outside of the brainstem in the carotid and aortic arteries, serve as the body's back-up respiratory drive by responding to low levels of O_2. This secondary mechanism is often referred to in COPD patients as a "hypoxic drive" since it takes over as the primary respiratory stimulation after the central chemoreceptors grow numb to chronically elevated $PaCO_2$.

Neurophysiology

Neurophysiology is a medical specialty that focuses on the relationship between the brain and the peripheral nervous system. As its name implies, neurophysiology is in many ways a melding of neurology, which is the study of the human brain and its functions, and physiology, which is the study of the sum of the body's parts and how they interrelate. Neurophysiologists examine the many ways in which brain activities impact nervous system activities. Much of the field's work is investigative, with doctors seeking to understand the origins of and best treatments for a variety of neurological disorders.

There are two parts to the human nervous system: the central nervous system, which is the brain and spinal cord, and the peripheral nervous system, which is the network of nerves that extends throughout the entire body. Nerves are responsible for sensitivity and feeling, but also muscle health and control. Neurophysiology examines the relationship between the two systems in causing degenerative diseases like multiple sclerosis and Parkinson's disease, as well as neurological disorders like epilepsy.

All parts of the body are ultimately controlled in the brain, but the brain, as a part of the nervous system, plays a unique role in nerve management. Neurophysiology tries to connect the role of the brain as nervous system controller with its role as nervous system member to better understand how nervous system problems happen and why. Doctors in the field will use tools and tests like electroencephalography and electromyography to study the ways in which affected nerves communicate with the brain. They uses this data to assess the general functioning of the nervous system as a whole, and to identify the roots of failures and problems.

Renal Physiology

Renal physiology is a discipline which involves the study of the function of the kidneys. Nephrologists, doctors who specialize in the kidneys, study renal physiology during their time in medical school, and an understanding of kidney function is also important to many nurses and doctors. Knowing how the kidneys work can help people identify problems with the kidney and address kidney function issues in their early stages.

Kidney physiology, as this field is also known, includes the study of all of the functions of the kidneys, from the time that fluids enter the kidneys to the moment that they are expressed. The kidneys use filtration, absorption, and secretion to manage a variety of systems within the body. They regulate the body's blood pressure, fluid balance, and balance of salts, and the kidneys also produce hormones which trigger various physiological responses.

The study of renal physiology usually focuses on the nephron, the smallest individual unit in the kidney. Each kidney contains numerous nephrons, with each nephron acting as a complete filtration and processing system which contributes to overall kidney function. As fluids pass along the nephron, the kidneys can selectively absorb or secrete, depending on the needs of the body. Someone who is dehydrated, for example, will have nephrons which retain water for the body, rather than allowing it to pass by, while someone who has just consumed a lot of water will have kidneys which retain salts to keep the balance of salts in the body stable.

Musculoskeletal Physiology

Musculoskeletal System

The musculoskeletal system consists of bones of the skeleton, the joints and the skeletal muscles. It provides form, support, stability, and movement to the body.

The musculoskeletal system's functions include supporting the body, allowing motion, and protecting vital organs. The skeletal also acts as the main storage system for calcium and phosphorus. Further, it contains important components of the hematopoietic system.

Bones are connected to other bones and muscle by tendons and ligaments. Bones provide stability to the body. Muscles hold the bones in place and also help in their movement. Different bones are connected by joints for producing motion. Cartilage prevents bones from rubbing directly onto each other. Muscles contract to move the bone attached at the joint.

Many diseases and disorders adversely affect the functioning of the musculoskeletal system. Some of these diseases may be difficult to diagnose due to the close relation of the musculoskeletal system with other organ systems.

Skeleton

- Bone
  - Types of bones
  - Bone structure
    - General structure of a long bone
    - Structure of short, irregular, flat and sesamoid bones
  - Microscopic structure of bone
    - Compact (cortical) bone

- ◇ Cancellous (trabecular, spongy) bone
- ◇ Bone cells
  - o Development of bone tissue (osteogenesis or ossification)
  - o Functions of bone
- Axial skeleton
  - o Skull
    - ◇ Cranium
    - ◇ Face
    - ◇ Sinuses
    - ◇ Fontanells of the skull
  - o Vertebral column
    - ◇ Characteristics of a typical vertebra
    - ◇ Special features of vertebra in different parts of the vertebral column
    - ◇ Features of the vertebral column
    - ◇ Functions of the vertebral column
  - o Thoracic cage
- Appendicular skeleton
  - o Shoulder girdle and upper limb
  - o Pelvic girdle and lower limb
- Healing of bones
  - o Factors that delay healing of fractures
  - o Complications of fractures
- Diseases of bones
  - o Osteoporosis
  - o Paget's disease
  - o Rickets and osteomalacia
  - o Infection of bones
  - o Osteomyelitis
  - o Developmental abnormalities of bone

     ○ Tumors of bone

       ◇ Benign tumors

       ◇ Malignant tumors

Joints

- Types of joint
  - Fibrous or fixed joints
  - Cartilaginous or slightly movable joints
  - Synovial or freely movable joints
    - ◇ Characteristics of a synovial joint
- Main synovial joints of the limbs
  - Shoulder joint
    - ◇ Muscles and movements
  - Elbow joint
    - ◇ Muscles and movements
  - Proximal and distal radioulnar joints
    - ◇ Muscles and movements
  - Wrist joints
    - ◇ Muscles and movements
  - Joints of the hands and fingers
  - Hip joint
    - ◇ Muscles and movements
  - Knee joint
    - ◇ Muscles and movements
  - Ankle joint
    - ◇ Muscles and movements
  - Joints of the foot and toes
- Disorders of joints
  - Inflammatory diseases of joints (arthritis)

- ◇ Rheumatoid arthritis (RA, rheumatoid disease)
- ◇ Other types of polyarthritis
- ◇ Infective arthritis
- o Traumatic injury to joints
  - ◇ Sprains, strains and dislocations
  - ◇ Penetrating injuries
- o Osteoarthritis (osteoarthrosos, OA)
  - ◇ Primary osteoarthritis
  - ◇ Secondary osteoarthritis
- o Gout
- o Connective tissue diseases
- o Carpal tunnel syndrome

Muscular System

- Muscles of the face and neck
  - o Muscles of the face
  - o Muscles of the neck
- Muscles of the back
- Muscles of the abdominal wall
  - o Functions
  - o Iguinal canal
- Muscles of the pelvic floor
  - o Functions
- Healing of muscle
- Repair of nerves supplying muscles
- Diseases of muscles
  - o Myasthenia gravis
  - o Myopathies
    - ◇ Muscular dystrophies

- Duchenne muscular dystrophy
- Facioscapulohumeral dystrophy
- Myotonic dystrophy
  - Crush syndrome.

References

- What-is-physiology, physiology-of-organisms: biology.cam.ac.uk, Retrieved 13 April 2018
- What-is-cardiovascular-physiology: wisegeek.com, Retrieved 23 April 2018
- Respiratory-system-physiology, human-physiology: jonathandownham.com, Retrieved 13 May 2018
- Physiology-respiration-10480389: emsworld.com, Retrieved 13 June 2018
- What-is-neurophysiology: wisegeek.com, Retrieved 30 March 2018
- What-is-renal-physiology: wisegeek.com, Retrieved 19 April 2018
- Musculoskeletal-system: anaphy.com, Retrieved 28 June 2018

Cell: Structure and Functions

The cell is the fundamental functional, structural and biological unit of living organisms. It consists of the cytoplasm, which is enclosed within a membrane, containing biomolecules such as nucleic acids and proteins. This chapter discusses in detail the various elements of cell structure and their functions. It includes topics such as cell nucleus, cytoplasm, cell junction, cell membrane and cell metabolism.

Cell

A cell is the basic unit of life as we know it. It is the smallest unit capable of independent reproduction. Robert Hooke suggested the name 'cell' in 1665, from the Latin "cella" meaning storeroom or chamber, after using a very early microscope to look at a piece of cork.

It is also said that he thought that the rectangular chambers looked like the cells in some monasteries.

Physically cells always have a boundary membrane, a little like a polythene bag encloses contents within it. Inside the space limited by the membrane there is a remarkable chemical processing unit.

From the point of view of cell structure biologists divide organisms into two groups, the bacteria (the prokaryotes), and all other animals and plants (the eukaryotes).

In bacteria chemical reactions take place almost anywhere within the cell. Bacteria contain genetic information in the form of DNA but it is not confined within a sac called a nucleus.

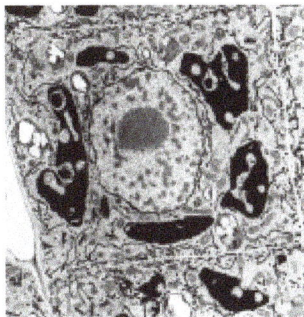

The main part of this image shows a root tip meristem cell from pea root. The black (osmium) staining shows the nuclear envelope, endoplasmic reticulum, Golgi apparatus and vacuoles.

In higher animals and plants specific functions are carried out by specialised structures. Collectively these are called organelles and include structures that contain the construction and operating plans of the cell (the nucleus), protein manufacturing areas (ribosomes), energy conversion units (mitochondria) and protein modifying and fat production areas (endoplasmic reticulum).

Additionally in plants there are light energy absorbers and converters (chloroplasts). Chloroplasts are almost unique in their capacity to convert sunlight energy into carbohydrate.

Cells also contain an elaborate transport network of filaments and fibres (the cytoskeleton) and a liquid (cytosol).

On the outer surface of a cell there can be a sticky material called extracellular matrix. This is proving to be very important to the cells it surrounds. Some animal cells produce bone and cartilage. Plant and animal cells have many features in common but plant cells also have a distinct rigid cell wall. Many plant cells also have large fluid filled sacs called vacuoles and some contain types of thickening that give plants rigidity and wood its unique strength.

Such is the efficiency of the cell that the main simple basic structure and function has been conserved during evolution and dispersal since cells started to form about 3.5 billion years ago.

The capacity and productivity of cells is truly amazing. In bacteria for example all the instructions come from a single closed loop of DNA. Each cell can divide in 20 minutes and given suitable conditions can keep dividing to produce 5 billion cells in eleven hours. Cells of this type produce some 400 different proteins and these are produced by enzyme assisted chemical reactions working at the rate of 100 times a second. This is why diseases such as meningitis and food poisoning can attack a person so quickly.

There is no such thing as a typical cell but most cells have chemical and structural features in common.

This is very important from the point of view of cell and molecular biology. It means that biologists can work on a cell from a mouse and be reasonably certain that the same processes will occur in a similar cell in a lion, a human or a fruit fly. This is possible because all cells are thought to have arisen from a common ancestor.

Many different types of plant and animal cells have evolved. In humans, there are about 200 different types but within cells there only about 20 different structures or organelles.

Many cells carry out specialized functions; this is what makes them different. The specialization of cells depends almost always on the exaggeration of properties common to cells. Cells lining the intestine for example have extended cell walls that increase the amount of surface area that is available to absorb food.

Nerve cells can be very long, extending for example in humans from the base of the spine to the foot.

Cells in heart muscle process a lot of energy and this is carried out by the high number of mitochondria found in these cells. At a molecular level however all cells resemble one another.

Cells vary greatly in their relative size although similar cells tend to be of similar size. Unfortunately most cells cannot be seen without a microscope.

The eggs of frogs and birds are large but they are made of a cell and a very large food store linked together. In relative size it has been suggested that the difference between the size of a bacterial cell and the egg of a frog would be the difference between a person and a frog's egg half a mile in diameter.

Cells are remarkable structures and in addition to facts mentioned already, they are able to communicate with each other receiving and rejecting messages.

Function of Cells

Scientists define seven functions that must be fulfilled by a living organism. These are:

1. A living thing must respond to changes in its environment.
2. A living thing must grow and develop across its lifespan.
3. A living thing must be able to reproduce, or make copies of itself.
4. A living thing must have metabolism.
5. A living thing must maintain homeostasis, or keep its internal environment the same regardless of outside changes.
6. A living thing must be made of cells.
7. A living thing must pass on traits to its offspring.

It is the biology of cells which enables living things to perform all of these functions.

Working of Cells

In order to accomplish them, they must have:

- A cell membrane that separates the inside of the cell from the outside. By concentrating the chemical reactions of life inside a small area within a membrane, cells allow the reactions of life to proceed much faster than they otherwise would.

- Genetic material, which is capable of passing on traits to the cell's offspring. In order to reproduce, organisms must ensure that their offspring have all the information that they need to be able to carry out all the functions of life. All modern cells accomplish this using DNA, whose base-pairing properties allow cells to make accurate copies of a cell's "blueprints" and "operating system." Some scientists think that the first cells might have used RNA instead.

- Proteins that perform a wide variety of structural, metabolic, and reproductive functions. There are countless different functions that cells must perform to obtain energy and reproduce. Depending on the cell, examples of these functions can include photosynthesis, breaking down sugar, locomotion, copying its own DNA, allowing certain substances to pass through the cell membrane while keeping others out, etc.

Proteins are made of amino acids, which are like the "Legos" of biochemistry. Amino acids come in different sizes, different shapes, and with different properties such as polarity, ionic charge, and hydrophobicity.

By putting amino acids together based on the instructions in their genetic material, cells can create biochemical machinery to perform almost any function.

Some scientists think that the first cells might have used RNA to accomplish some vital functions, and then moved to much more versatile amino acids to do the job as the result of a mutation.

Other Components of Cells

A cell's cytoplasm is home to numerous functional and structural elements. These elements exist in the form of molecules and organelles — picture them as the tools, appliances, and inner rooms of the cell. Major classes of intracellular organic molecules include nucleic acids, proteins, carbohydrates, and lipids, all of which are essential to the cell's functions.

Nucleic acids are the molecules that contain and help express a cell's genetic code. There are two major classes of nucleic acids: deoxyribonucleic acid (DNA) and ribonucleic acid (RNA). DNA is the molecule that contains all of the information required to build and maintain the cell; RNA has several roles associated with expression of the information stored in DNA. Of course, nucleic acids alone aren't responsible for the preservation and expression of genetic material: Cells also use proteins to help replicate the genome and accomplish the profound structural changes that underlie cell division.

Proteins are a second type of intracellular organic molecule. These substances are made from chains of smaller molecules called amino acids, and they serve a variety of functions in the cell, both catalytic and structural. For example, proteins called enzymes convert cellular molecules (whether proteins, carbohydrates, lipids, or nucleic acids) into other forms that might help a cell meet its energy needs, build support structures, or pump out wastes.

Carbohydrates, the starches and sugars in cells, are another important type of organic molecule. Simple carbohydrates are used for the cell's immediate energy demands, whereas complex carbohydrates serve as intracellular energy stores. Complex carbohydrates are also found on a cell's surface, where they play a crucial role in cell recognition.

Finally, lipids or fat molecules are components of cell membranes — both the plasma membrane and various intracellular membranes. They are also involved in energy storage, as well as relaying signals within cells and from the bloodstream to a cell's interior.

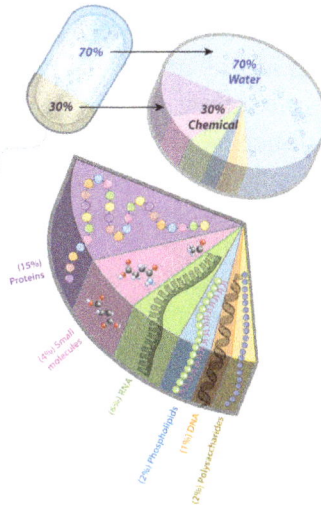

Figure: Composition of a bacterial cell

Most of a cell is water (70%). The remaining 30% contains varying proportions of structural and functional molecules. Some cells also feature orderly arrangements of molecules called organelles. Similar to the rooms in a house, these structures are partitioned off from the rest of a cell's interior by their own intracellular membrane. Organelles contain highly technical equipment required for specific jobs within the cell. One example is the mitochondrion — commonly known as the cell's "power plant" — which is the organelle that holds and maintains the machinery involved in energy-producing chemical reactions.

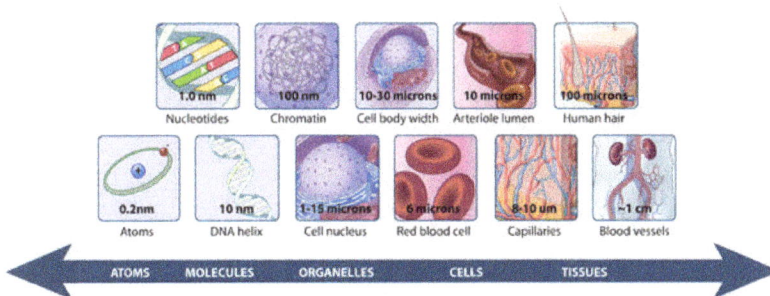

Figure: Relative scale of biological molecules and structures

Cells can vary between 1 micrometer (μm) and hundreds of micrometers in diameter.

Within a cell, a DNA double helix is approximately 10 nanometers (nm) wide, whereas the cellular organelle called a nucleus that encloses this DNA can be approximately 1000 times bigger (about 10 μm). See how cells compare along a relative scale axis with other molecules, tissues, and biological structures (blue arrow at bottom). Note that a micrometer (μm) is also known as a micron.

Different Categories of Cells

Rather than grouping cells by their size or shape, scientists typically categorize them by how their genetic material is packaged. If the DNA within a cell is not separated from the cytoplasm, then that cell is a prokaryote. All known prokaryotes, such as bacteria and archaea, are single cells. In contrast, if the DNA is partitioned off in its own membrane-bound room called the nucleus, then that cell is a eukaryote. Some eukaryotes, like amoebae, are free-living, single-celled entities. Other eukaryotic cells are part of multicellular organisms. For instance, all plants and animals are made of eukaryotic cells — sometimes even trillions of them.

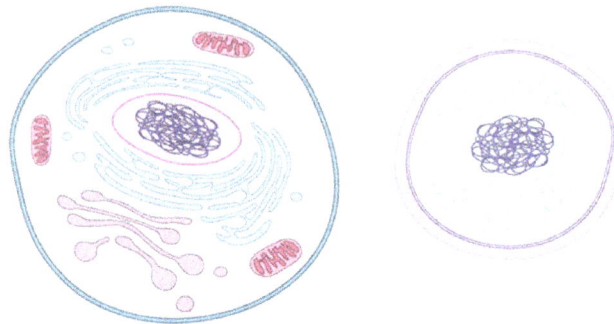

Figure: Comparing basic eukaryotic and prokaryotic differences

A eukaryotic cell (left) has membrane-enclosed DNA, which forms a structure called the nucleus (located at center of the eukaryotic cell; note the purple DNA enclosed in the pink nucleus). A typical eukaryotic cell also has additional membrane-bound organelles of varying shapes and sizes. In contrast, a prokaryotic cell (right) does not have membrane-bound DNA and also lacks other membrane-bound organelles as well.

Cell Nucleus

The nucleus is a highly specialized organelle that serves as the information processing and administrative center of the cell. This organelle has two major functions: it stores the cell's hereditary material, or DNA, and it coordinates the cell's activities, which include growth, intermediary metabolism, protein synthesis, and reproduction (cell division).

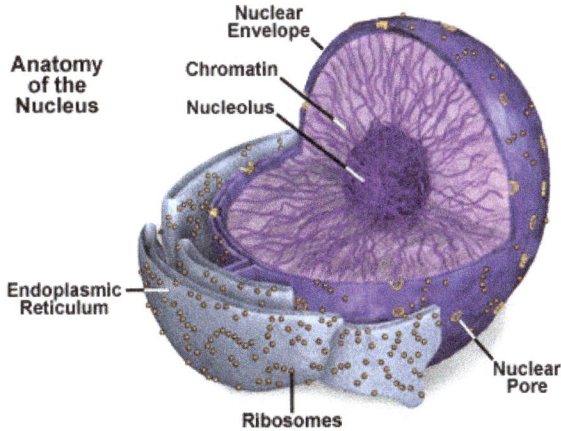

Anatomy of the Nucleus

Nuclear Envelope, Chromatin, Nucleolus, Endoplasmic Reticulum, Nuclear Pore, Ribosomes

Only the cells of advanced organisms, known as eukaryotes, have a nucleus. Generally there is only one nucleus per cell, but there are exceptions, such as the cells of slime molds and the Siphonales group of algae. Simpler one-celled organisms (prokaryotes), like the bacteria and cyanobacteria, don't have a nucleus. In these organisms, all of the cell's information and administrative functions are dispersed throughout the cytoplasm.

The spherical nucleus typically occupies about 10 percent of a eukaryotic cell's volume, making it one of the cell's most prominent features. A double-layered membrane, the nuclear envelope, separates the contents of the nucleus from the cellular cytoplasm. The envelope is riddled with holes called nuclear pores that allow specific types and sizes of molecules to pass back and forth between the nucleus and the cytoplasm. It is also attached to a network of tubules and sacs, called the endoplasmic reticulum, where protein synthesis occurs, and is usually studded with ribosomes.

The semifluid matrix found inside the nucleus is called nucleoplasm. Within the nucleoplasm, most of the nuclear material consists of chromatin, the less condensed form of the cell's DNA that organizes to form chromosomes during mitosis or cell division. The nucleus also contains one or more nucleoli, organelles that synthesize protein-producing macromolecular assemblies called ribosomes, and a variety of other smaller components, such as Cajal bodies, GEMS (Gemini of coiled bodies), and interchromatin granule clusters.

Chromatin and chromosomes: Packed inside the nucleus of every human cell is nearly 6 feet of DNA, which is divided into 46 individual molecules, one for each chromosome and each about 1.5 inches long. Packing all this material into a microscopic cell nucleus is an extraordinary feat of packaging. For DNA to function, it can't be crammed into the nucleus like a ball of string. Instead, it is combined with proteins and organized into a precise, compact structure, a dense string-like fiber called chromatin.

Nucleolus: The nucleolus is a membrane-less organelle within the nucleus that manufactures ribosomes, the cell's protein-producing structures. Through the microscope, the nucleolus looks like a large dark spot within the nucleus. A nucleus may contain up to four nucleoli, but within each species the number of nucleoli is fixed. After a cell divides, a nucleolus is formed when chromosomes are brought together into nuclear organizing regions. During cell division, the nucleolus disappears. Some studies suggest that the nucleolus may be involved with cellular aging and, therefore, may affect the senescence of an organism.

Nuclear envelope: The nuclear envelope is a double-layered membrane that encloses the contents of the nucleus during most of the cell's lifecycle. The space between the layers is called the perinuclear space and appears to connect with the rough endoplasmic reticulum. The envelope is perforated with tiny holes called nuclear pores. These pores regulate the passage of molecules between the nucleus and cytoplasm, permitting some to pass through the membrane, but not others. The inner surface has a protein lining called the nuclear lamina, which binds to chromatin and other nuclear components. During mitosis, or cell division, the nuclear envelope disintegrates, but reforms as the two cells complete their formation and the chromatin begins to unravel and disperse.

Nuclear pores: The nuclear envelope is perforated with holes called nuclear pores. These pores regulate the passage of molecules between the nucleus and cytoplasm, permitting some to pass through the membrane, but not others. Building blocks for building DNA and RNA are allowed into the nucleus as well as molecules that provide the energy for constructing genetic material.

Cytoplasm

Cytoplasm refers to the fluid that fills the cell, which includes the cytosol along with filaments, proteins, ions and macromolecular structures as well as the organelles suspended in the cytosol.

In eukaryotic cells, cytoplasm refers to the contents of the cell with the exception of the nucleus. Eukaryotes have elaborate mechanisms for maintaining a distinct nuclear compartment separate from the cytoplasm. Active transport is involved in the creation of these subcellular structures and for maintaining homeostasis with the cytoplasm. For prokaryotic cells, since they do not have a defined nuclear membrane, the cytoplasm also contains the cell's primary genetic material. These cells are usually smaller in comparison to eukaryotes, and have a simpler internal organization of the cytoplasm.

Structure of Cytoplasm

The cytoplasm is unusual because it is unlike any other fluid found in the physical world. Liquids that are studied to understand diffusion usually contain a few solutes in an aqueous environment. However, the cytoplasm is a complex and crowded system containing a wide range of particles – from ions and small molecules, to proteins as well as giant multi protein complexes and organelles. These constituents are moved across the cell depending on the requirements of the cell along an elaborate cytoskeleton with the help of specialized motor proteins. The movement of such large particles also changes the physical properties of the cytosol.

The physical nature of the cytoplasm is variable. Sometimes, there is quick diffusion across the cell, making the cytoplasm resemble a colloidal solution. At other times, it appears to take on the properties of a gel-like or glass-like substance. It is said to have the properties of viscous as well as elastic materials – capable of deforming slowly under external force in addition to regaining its original shape with minimal loss of energy. Parts of the cytoplasm close to the plasma membrane are also 'stiffer' while the regions near the interior resemble free flowing liquids. These changes in the cytoplasm appear to be dependent on the metabolic processes within the cell and play an important role in carrying out specific functions and protecting the cell from stressors.

The cytoplasm can be divided into three components:

1. The cytoskeleton with its associated motor proteins

2. Organelles and other large multi-protein complexes

3. Cytoplasmic inclusions and dissolved solutes

Cytoskeleton and Motor Proteins

The basic shape of the cell is provided by its cytoskeleton formed primarily by three types of polymers – actin filaments, microtubules and intermediate filaments.

Actin filaments or microfilaments are 7 nm in width and are made of double stranded polymers of F-actin. These filaments are associated with a number of other proteins that help in filament assembly and are also involved in anchoring them close to the plasma membrane. This cytoplasmic location helps the microfilaments become involved in rapid responses to signal molecules from the extracellular environment and produce cellular responses through signal transduction or chemotaxis. In addition, myosin, an ATP-based motor protein transmits cargo and vesicles along the microfilament and is also involved in muscle contraction.

Microtubules are polymers of α and β tubulin, which form a hollow tube by the lateral association of 13 protofilaments. Each protofilament is a polymer of alternating α and β tubulin molecules. The inner diameter of a microtubule is 12 nm and its outer diameter is 24 nm.

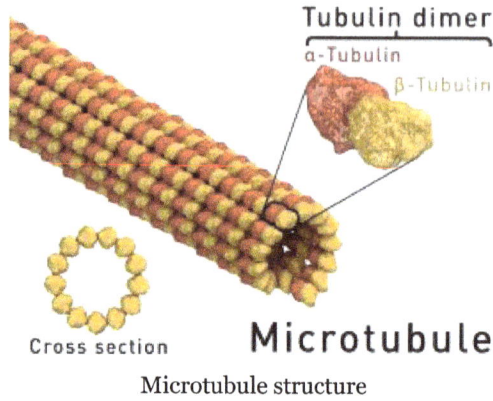
Microtubule structure

Microtubules radiate towards the periphery of the cell from microtubule organizing centers (MTOCs) located close to the nucleus, and provide structure and shape to the cell.

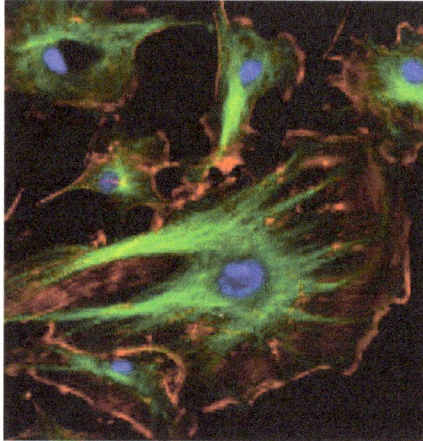
Fluorescent Cells

This image shows the nucleus in blue, the actin filaments on the cell periphery are labeled red and the extensive microtubule network is marked green. The cytoplasm undergoes rapid reorganization during cell division with microtubules forming the spindle, which binds to chromosomes and segregates them into two daughter cells.

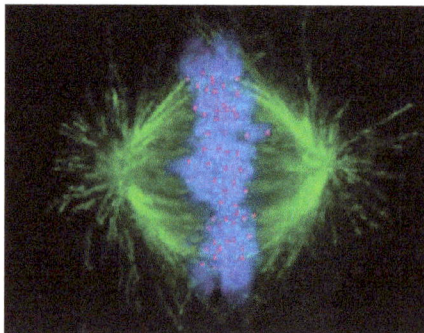
Kinetochore

Similar to the previous image, chromosomes are stained blue and microtubules are green. Tiny red dots are kinetochores.

Microtubules are involved in cytoplasmic transport, chromosome segregation and in forming structures such as cilia and flagella for cellular movement.

Intermediate filaments are larger than microfilaments but smaller than microtubules and are formed by a group of proteins that share structural features. Though they are not involved in cell motility, they are important for cells to come together as tissues and to remain anchored to the extracellular matrix.

Organelles and Multi-protein Complexes

Most eukaryotic cells have a number of organelles that provide compartments within the cytoplasm for specialized microenvironments. For instance, lysosomes contain a number of hydrolases in an acidic environment that is ideal for their enzymatic activity. These hydrolases are actively transported into the lysosome after being synthesized in the cytoplasm. Mitochondria, while containing their own genome, also need many enzymes synthesized in the cytosol, which are then selectively moved into the organelle. These organelles are placed in specific locations due to the physical gel-like nature of the cytoplasm and by anchoring to the cytoskeleton.

In addition, the cytoplasm also plays host to multi-protein complexes like the proteasome and ribosomes. Ribosomes are large complexes of RNA and protein that are important for the translation of mRNA code into amino acid sequences of proteins. Proteasomes are giant molecular structures about 20,000 kilodaltons in mass and 15 nm in diameter. Proteasomes are important for targeted destruction of proteins that are no longer needed by the cell.

Cytoplasmic Inclusions

Cytoplasmic inclusions can include a wide range of biochemicals – from small crystals of proteins, to pigments, carbohydrates and fats. All cells, especially in tissue like the adipose, contain droplets of lipids in their triglyceride form. These are used to create cellular membranes and are an excellent energy store. Lipids can generate twice as many ATP molecules per gram when compared to carbohydrates. However, the process of releasing this energy from triglycerides in intensive in oxygen consumption and therefore the cell also contains stores of glycogen as cytoplasmic inclusions. Glycogen inclusions are particularly important in cells like the skeletal and cardiac muscle cells where there can be a sudden increase in demand for glucose. Glycogen can be quickly broken down into individual molecules of glucose and used in cellular respiration before the cell can obtain more glucose reserves from the body.

Crystals are another type of cytoplasmic inclusion found in many cells, and have special function in cells of the inner ear (maintaining balance). Presence of crystals in cells of the testis appears to be linked with morbidity and infertility. Finally, the cytoplasm also contains pigments such as melanin, which lead to the pigmented cells of the skin. These pigments protect the cell and internal body structures from the deleterious effects of ultraviolet radiation. Pigments are also prominent in the cells of the iris that surround the pupil of the eye.

Each of these components affects the functioning of the cytoplasm in different ways, making it a dynamic region that plays a role in, and is influenced by the cell's overall metabolic activity.

Functions of Cytoplasm

The cytoplasm is the site for most of the enzymatic reactions and metabolic activity of the cell. Cellular respiration begins in the cytoplasm with anaerobic respiration or glycolysis. This reaction provides the intermediates that are used by the mitochondria to generate ATP. In addition, the translation of mRNA into proteins on ribosomes also occurs mostly in the cytoplasm. Some of it happens on free ribosomes suspended in the cytosol while the rest happens on ribosomes anchored on the endoplasmic reticulum.

The cytoplasm also contains the monomers that go on to generate the cytoskeleton. The cytoskeleton, in addition to being important for the normal activities of the cell is crucial for cells that have a specialized shape. For instance, neurons with their long axons need the presence of intermediate filaments, microtubules, and actin filaments in order to provide a rigid framework for the action potential to be transmitted to the next cell. Additionally, some epithelial cells contain small cilia or flagella to move the cell or remove foreign particles through coordinated activity of cytoplasmic extrusions formed through the cytoskeleton.

The cytoplasm also plays a role in creating order within the cell with specific locations for different organelles. For instance, the nucleus is usually seen towards the center of the cell, with a centrosome nearby. The extensive endoplasmic reticulum and Golgi network are also placed in relation to the nucleus, with the vesicles radiating out towards the plasma membrane.

Cytoplasmic Streaming

Movement within the cytoplasm also occurs in bulk, through the directed movement of cytosol around the nucleus or vacuole. This is particularly important in large single celled organisms such as some species of green algae, which can be nearly 10 cm in length. Cytoplasmic streaming is also important for positioning chloroplasts close to the plasma membrane to optimize photosynthesis and for distributing nutrients through the entire cell. In some cells, such as mouse oocytes, cytoplasmic streaming is expected to have a role in the formation of cellular sub-compartments and in organelle positioning as well.

Cytoplasmic Inheritance

The cytoplasm plays hosts to two organelles that contain their own genomes – the chloroplast and mitochondria. These organelles are inherited directly from the mother through the oocyte and therefore constitute genes that are inherited outside the nucleus. These organelles replicate independent of the nucleus and respond to the needs of the cell. Cytoplasmic or extra nuclear inheritance, therefore, forms an unbroken genetic line that has not undergone mixing or recombination with the male parent.

Cell Membrane

Cell Membrane Structure

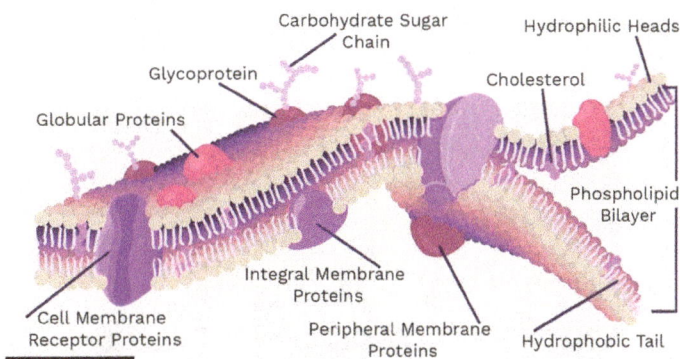

The cell membrane (plasma membrane) is a thin semi-permeable membrane that surrounds the cytoplasm of a cell. Its function is to protect the integrity of the interior of the cell by allowing certain substances into the cell, while keeping other substances out. It also serves as a base of attachment for the cytoskeleton in some organisms and the cell wall in others. Thus, the cell membrane also serves to help support the cell and help maintain its shape.

Another function of the membrane is to regulate cell growth through the balance of endocytosis and exocytosis. In endocytosis, lipids and proteins are removed from the cell membrane as substances are internalized. In exocytosis, vesicles containing lipids and proteins fuse with the cell membrane increasing cell size. Animal cells, plant cells, prokaryotic cells, and fungal cells have plasma membranes. Internal organelles are also encased by membranes.

Cell Membrane Structure

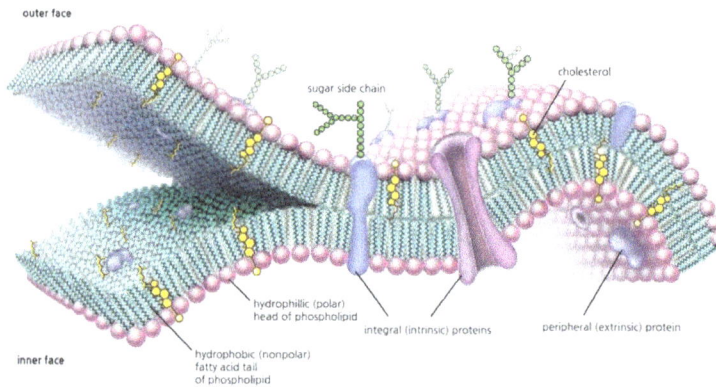

The cell membrane is primarily composed of a mix of proteins and lipids. Depending on the membrane's location and role in the body, lipids can make up anywhere from 20 to 80 percent of the membrane, with the remainder being proteins. While lipids help to give membranes their flexibility, proteins monitor and maintain the cell's chemical climate and assist in the transfer of molecules across the membrane.

Cell Membrane Lipids

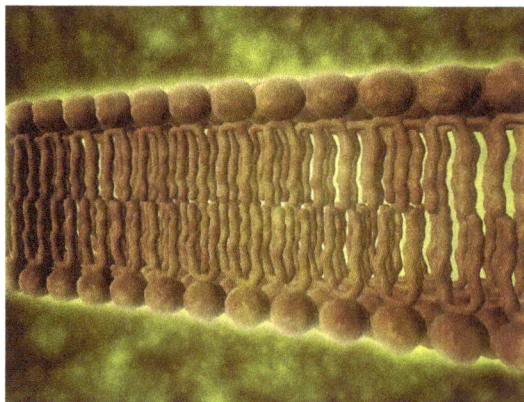

Phospholipids are a major component of cell membranes. Phospholipids form a lipid bilayer in which their hydrophilic (attracted to water) head areas spontaneously arrange to face the aqueous cytosol and the extracellular fluid, while their hydrophobic (repelled by water) tail areas face away from the cytosol and extracellular fluid. The lipid bilayer is semi-permeable, allowing only certain molecules to diffuse across the membrane.

Cholesterol is another lipid component of animal cell membranes. Cholesterol molecules are selectively dispersed between membrane phospholipids. This helps to keep cell membranes from becoming stiff by preventing phospholipids from being too closely packed together. Cholesterol is not found in the membranes of plant cells.

Glycolipids are located on cell membrane surfaces and have a carbohydrate sugar chain attached to them. They help the cell to recognize other cells of the body.

Cell Membrane Proteins

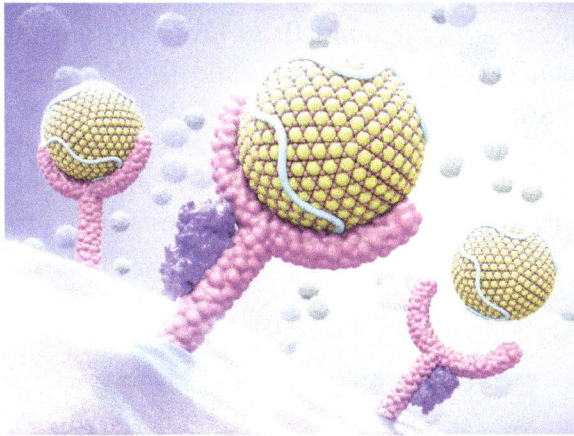

The cell membrane contains two types of associated proteins. Peripheral membrane proteins are exterior to and connected to the membrane by interactions with other proteins. Integral membrane proteins are inserted into the membrane and most pass through the membrane. Portions of these transmembrane proteins are exposed on both sides of the membrane. Cell membrane proteins have a number of different functions.

Structural proteins help to give the cell support and shape.

Cell membrane receptor proteins help cells communicate with their external environment through the use of hormones, neurotransmitters, and other signaling molecules.

Transport proteins, such as globular proteins, transport molecules across cell membranes through facilitated diffusion.

Glycoproteins have a carbohydrate chain attached to them. They are embedded in the cell membrane and help in cell to cell communications and molecule transport across the membrane.

Organelle Membranes

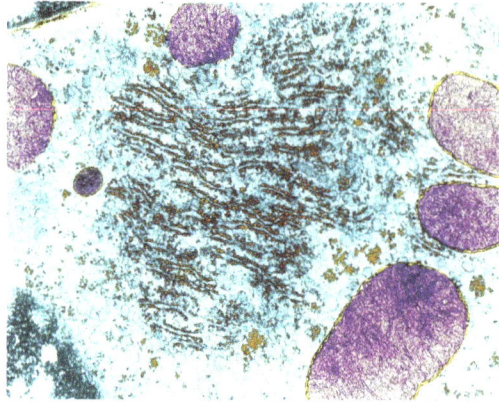

Some cell organelles are also surrounded by protective membranes. The nucleus, endoplasmic reticulum, vacuoles, lysosomes, and Golgi apparatus are examples of membrane-bound organelles. Mitochondria and chloroplasts are bound by a double membrane. The membranes of the different organelles vary in molecular composition and are well suited for the functions they perform. Organelle membranes are important to several vital cell functions including protein synthesis, lipid production, and cellular respiration.

Eukaryotic Cell Structures

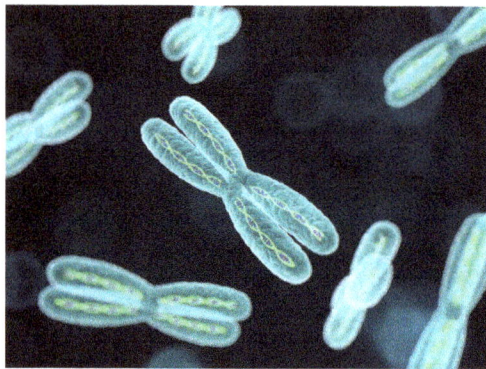

The cell membrane is only one component of a cell. The following cell structures can also be found in a typical animal eukaryotic cell:

- Centrioles - help to organize the assembly of microtubules.

- Chromosomes - house cellular DNA.

- Cilia and Flagella - aid in cellular locomotion.

- Endoplasmic Reticulum - synthesizes carbohydrates and lipids.

- Golgi Apparatus - manufactures, stores and ships certain cellular products.

- Lysosomes - digest cellular macromolecules.

- Mitochondria - provide energy for the cell.

- Nucleus - controls cell growth and reproduction.

- Peroxisomes - detoxify alcohol, form bile acid, and use oxygen to break down fats.

- Ribosomes - responsible for protein production via translation.

Cell Junction

Cell junctions can be divided into two types: those that link cells together, also called intercellular junctions (tight, gap, adherens, and desmosomal junctions), and those that link cells to the extracellular matrix (focal contacts/adhesion plaques and hemi desmosomes). These junctions play a prominent role in maintaining the integrity of tissues in multicellular organisms and some, if not all of them, are involved in signal transduction.

Intercellular junctions and hemidesmosomes were first identified in tissues examined by electron microscopy. In contrast, the focal contact was first observed in cultured cells in the light microscope by a technique called interference reflection. This procedure revealed specific sites where cells closely adhere to their substrate. These were called focal contacts or adhesion plaques.

Tight Junctions

Tight junctions are areas where the membranes of two adjacent cells join together to form a barrier. The cell membranes are connected by strands of transmembrane proteins such as claudins and occludins. Tight junctions bind cells together, prevent molecules from passing in between the cells, and also help to maintain the polarity of cells. They are only found in vertebrates, animals with a backbone and skeleton; invertebrates have septate junctions instead.

Function of Tight Junctions

Tight junctions have several different functions. Their most important functions are to help cells form a barrier that prevents molecules from getting through, and to stop proteins in the cell membrane from moving around. Tight junctions are often found at epithelial cells, which are cells that line the surface of the body and line body cavities. Not only do epithelial cells separate the body from the surrounding environment, they also separate surfaces within the body. Therefore, it is very important that the permeability of molecules through layers of epithelial cells is tightly controlled.

If molecules are blocked by tight junctions and physically unable to pass through the space in between cells, they must enter through other methods that involve entering the cells themselves. They could pass through special proteins in the cell membrane, or be engulfed by the cell through endocytosis. Using these methods, the cell has greater control over what materials it takes in and allows to pass through. However, in endothelial cells, certain proteins must be kept on certain sides of the cell. The apical, or outside layer, of the sheet of cells contains proteins that only let certain substances pass through. The basal, or inside layer, is where cells let molecules pass through them by expelling them from their membrane in a process called exocytosis. Exocytosis also relies on specific proteins in order to work correctly. Tight junctions keep the correct proteins on the correct sides of the cell in order for these functions to occur. This also helps maintain the polarity of cells.

Another function of tight junctions is simply to hold cells together. The branching protein strands of tight junctions link adjacent cells together tightly so that they form a sheet. These strands are anchored to microfilaments, part of the cell's cytoskeleton that is made up of long strands of actin proteins. Microfilaments are located inside the cell, so the combination of microfilaments and sealing strands anchors the cells together from the inside and the outside.

Structure of Tight Junctions

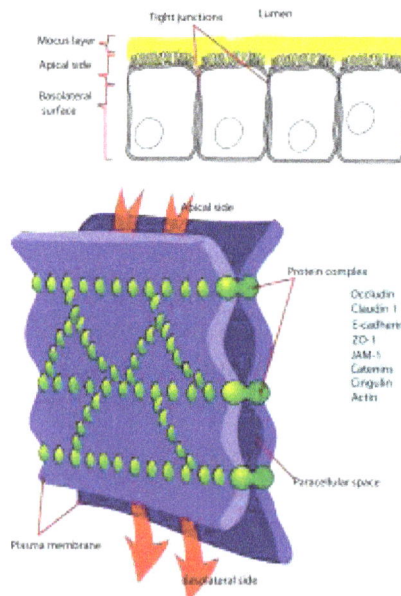

Figure: Tight junction between cells

Tight junctions are a branching network of protein strands on the surface of a cell that link with each other throughout the surface of the membrane. The strands are formed by transmembrane proteins on the surfaces of the cell membranes that are adjacent to each other.

There are around 40 different proteins at tight junctions. These proteins can be grouped into four main types. Transmembrane proteins are wedged in the middle of the cell membrane and are responsible for adhesion and permeability. Scaffolding proteins organize transmembrane proteins. Signaling proteins are responsible for forming the tight junction and regulating the barrier. Regulation proteins regulate what proteins are brought to the cell membrane in vesicles.

Claudins and occludins are the two main types of proteins present at tight junctions, and they are both transmembrane proteins. Claudins are important in forming tight junctions, while occludins play more of a role in keeping the tight junction stable and maintaining the barrier between cells that keeps unwanted molecules out.

Gap Junctions

Gap junctions are a type of cell junction in which adjacent cells are connected through protein channels. These channels connect the cytoplasm of each cell and allow molecules, ions, and electrical signals to pass between them. Gap junctions are found in between the vast majority of cells within the body because they are found between all cells that are directly touching other cells. Exceptions include cells that move around and do not usually come into close contact with other cells, such as sperm cells and red blood cells. Gap junctions are only found in animal cells; plant cells are connected by channels called plasmodesmata instead.

Function of Gap Junctions

The main function of gap junctions is to connect cells together so that molecules may pass from one cell to the other. This allows for cell-to-cell communication, and makes it so that molecules can directly enter neighboring cells without having to go through the extracellular fluid surrounding the cells. Gap junctions are especially important during embryonic development, a time when neighboring cells must communicate with each other in order for them to develop in the right place at the right time. If gap junctions are blocked, embryos cannot develop normally.

Gap junctions make cells chemically or electrically coupled. This means that the cells are linked together and can transfer molecules to each other for use in reactions. Electrical coupling occurs in the heart, where cells receive the signal to contract the heart muscle at the same time through gap junctions. It also occurs in neurons, which can be connected to each other by electrical synapses in addition to the well-known chemical synapses that neurotransmitters are released from.

When a cell starts to die from disease or injury, it sends out signals through its gap junctions. These signals can cause nearby cells to die even if they are not diseased or injured. This is called the "bystander effect", since the nearby cells are innocent bystanders that become victims. However, sometimes groups of adjacent cells need to die during development, so gap junctions facilitate this process. In addition, cells can also send therapeutic compounds to each other through gap junctions, and gap junctions are being researched as a method of therapeutic drug delivery.

Gap Junction Structure

In vertebrate cells, gap junctions are made up of connexin proteins. (The cells of invertebrates have gap junctions that are composed of innexin proteins, which are not related to connexin proteins but perform a similar function.) Groups of six connexins form a connexon, and two connexons are put together to form a channel that molecules can pass through. Other channels in gap junctions are made up of pannexin proteins. Relatively less is known about pannexins; they were originally thought only to form channels within a cell, not between cells. Hundreds of channels are found together at the site of a gap junction in what is known as a gap junction plaque. A plaque is a mass of proteins.

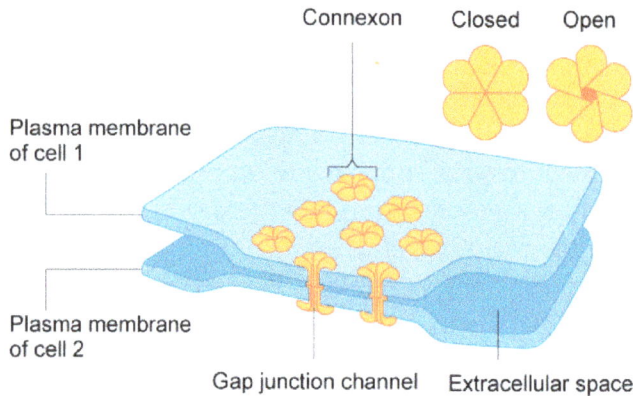

Figure: Channels at a gap junction.

Adherens Junctions

Adherens junctions provide strong mechanical attachments between adjacent cells through the linkage of cytoplasmic face with cytoskeleton.

Adherens junctions are also referred to as zonula adherens, intermediate junction, or as belt desmosomes. Zonula means small zone or belt-like, and adherens refers to adhesion (sticking together). As a result, the zonula adherens often runs like a belt around the entire cell in a continuous fashion, and it acts as an adhesion belt.

Location and Function

This type of cell junction is located right below tight junctions and provides a strong bond between the sides of adjacent epithelial cell membranes. While other junctions, like tight junctions, provide some support for and fusion of adjacent cells, their resistance to mechanical stress is relatively small compared to the much stronger adherens junctions.

Structure and Composition

The zonula adherens is composed of several different proteins:

- The actin microfilaments of the cytoskeleton (the internal skeleton of the cell).

- Anchor proteins, found inside each cell. These are called alpha-catenin, beta-catenin, gamma-catenin (aka plakoglobin), vinculin, and alpha-actinin. They link the actin microfilaments to the cadherins.

- Cadherins, namely E-cadherin. These are transmembrane adhesion proteins, whose main portions are located in the extracellular space.

The extracellular part of one cell's cadherin binds to the extracellular part of the adjacent cell's cadherin in the space between the two cells. Each cell's cadherin molecule also contains a tail that inserts itself inside its respective cell.

This intracellular (within the cell) tail then links up to catenin proteins to form the cadherin–catenin complex. This complex binds to vinculin and alpha-actinin; these two proteins are what link the cadherin–catenin complex to the cell's internal skeletal framework (the actin microfilaments).

The extracellular portions of the cadherin molecules of adjacent cells are bonded together by calcium ions (or another protein in some cases). This means that the functional as well as morphological integrity of the adherens junctions are calcium dependent. If you were to remove calcium from the equation, this type of cell junction would disintegrate as a result.

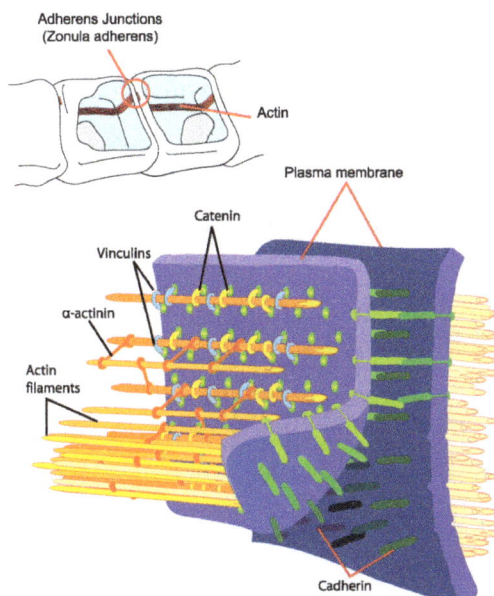

The structural proteins in an adherens junction: These are the principal interactions of structural proteins at a cadherin-based plasma membrane adherens junction. Actin filaments are associated with adherens junctions in addition to several other actin-binding proteins.

Cell Metabolism

The collection of chemical reactions in the body is usually referred to as the metabolism. This process is the sum of all chemical changes that take place within the cells in the body. During digestion, for example, cellular metabolism is what releases energy from nutrients. Cellular metabolism sustains life and allows cells to grow, develop, repair damage, and respond to environmental changes.

Cellular metabolism can break down organic matter, a process known as catabolism. Cellular metabolism can also produce substances, a process referred to as anabolism. To provide a more graspable example, breaking down food so the nutrients can be utilized is a catabolic reaction. The production of proteins from amino acids is an example of an anabolic reaction. In general, breaking down releases energy and building up consumes energy. Amino acids, carbohydrates, and lipids (often called fats) are vital for life. Metabolic reactions either produce these molecules during the construction of cells and tissue or digest them and use them as a source of energy.

Catabolic Cellular Metabolism

Catabolic metabolism breaks down complex organic molecules into more simple molecules. These exergonic reactions are characterized by the release of energy. Catabolism reduces protein, fat, and carbohydrates into amino acids, fatty acids, and simple sugars, respectively. The energy released from catabolic reactions drives anabolic reactions. It's a process that has three stages:

1. Breakdown of complex molecules into their basic building blocks.

2. Breakdown of the basic building blocks into even more simple metabolic intermediates.

3. "Combustion" of the acetyl groups of acetyl-coenzyme A by the citric acid cycle and oxidative phosphorylation to produce CO_2 and H_2O. In other words, energy is released.

Anabolic Cellular Metabolism

Whereas catabolic metabolism breaks down molecules into their constituents, anabolic metabolism combines simple substances into more complex substances. When the cells combine amino acids into proteins to produce cells or tissues, that's anabolism. Anabolic reactions are endergonic reactions, which means they use more energy than they produce.

Although catabolism and anabolism occur independently of each other, they are inextricably linked. Without cellular metabolism, the body's cells wouldn't be able to break down or synthesize the compounds needed for energy, growth, function, and healing.

Metabolic Pathways

Many of the molecular transformations that occur within cells require multiple steps to accomplish. Recall, for instance, that cells split one glucose molecule into two pyruvate molecules by way of a ten-step process called glycolysis. This coordinated series of chemical reactions is an example of a metabolic pathway in which the product of one reaction becomes the substrate for the next reaction. Consequently, the intermediate products of a metabolic pathway may be short-lived.

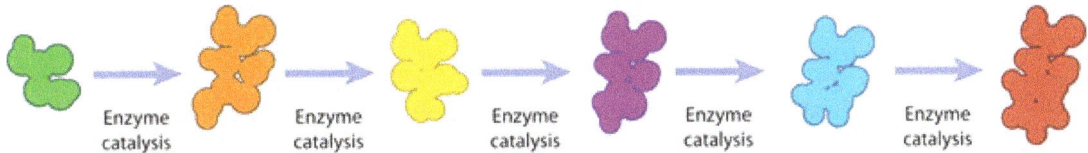

Enzyme catalysis   Enzyme catalysis   Enzyme catalysis   Enzyme catalysis   Enzyme catalysis

Figure: Reaction pathway

Enzymes can be involved at every step in a reaction pathway. At each step, the molecule is transformed into another form, due to the presence of a specific enzyme. Such a reaction pathway can create a new molecule (biosynthesis), or it can break down a molecule (degradation).

Sometimes, the enzymes involved in a particular metabolic pathway are physically connected, allowing the products of one reaction to be efficiently channeled to the next enzyme in the pathway. For example, pyruvate dehydrogenase is a complex of three different enzymes that catalyze the path from pyruvate (the end product of glycolysis) to acetyl CoA (the first substrate in the citric acid cycle). Within this complex, intermediate products are passed directly from one enzyme to the next.

Balance of Chemical Reactions in Cells

Cells are expert recyclers. They disassemble large molecules into simpler building blocks and then use those building blocks to create the new components they require. The breaking down of complex organic molecules occurs via catabolic pathways and usually involves the release of energy. Through catabolic pathways, polymers such as proteins, nucleic acids, and polysaccharides are reduced to their constituent parts: amino acids, nucleotides, and sugars, respectively. In contrast, the synthesis of new macromolecules occurs via anabolic pathways that require energy input.

Cells must balance their catabolic and anabolic pathways in order to control their levels of critical metabolites — those molecules created by enzymatic activity — and ensure that sufficient energy is available. For example, if supplies of glucose start to wane, as might happen in the case of starvation, cells will synthesize glucose from other materials or start sending fatty acids into the citric acid cycle to generate ATP. Conversely, in times of plenty, excess glucose is converted into storage forms, such as glycogen, starches, and fats.

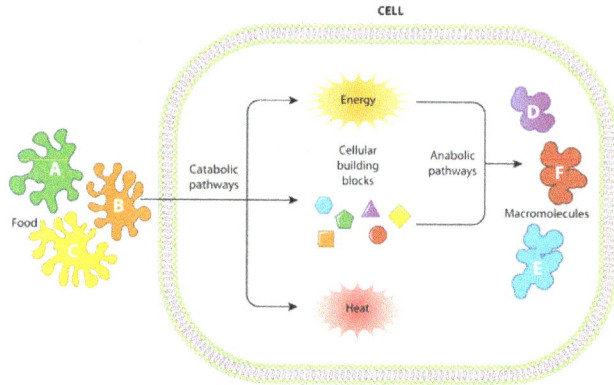

Figure: Catabolic and anabolic pathways in cell metabolism

Catabolic pathways involve the breakdown of nutrient molecules (Food: A, B, C) into usable forms (building blocks). In this process, energy is either stored in energy molecules for later use, or released as heat. Anabolic pathways then build new molecules out of the products of catabolism, and these pathways typically use energy. The new molecules built via anabolic pathways (macromolecules) are useful for building cell structures and maintaining the cell.

Chemical Reactions Management in Cells

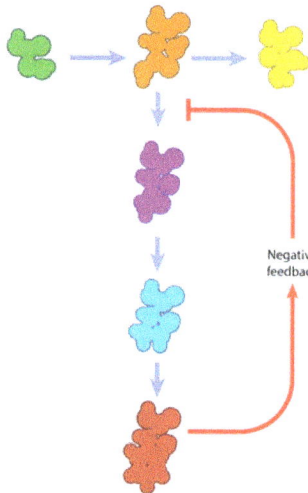

Figure: Feedback inhibition

When there is enough product at the end of a reaction pathway (red macromolecule), it can inhibit its own synthesis by interacting with enzymes in the synthesis pathway (red arrow).

Not only do cells need to balance catabolic and anabolic pathways, but they must also monitor the needs and surpluses of all their different metabolic pathways. In order to bolster a particular pathway, cells can increase the amount of a necessary (rate-limiting) enzyme or use activators to convert that enzyme into an active conformation. Conversely, to slow down or halt a pathway, cells can decrease the amount of an enzyme or use inhibitors to make the enzyme inactive.

Such up- and down-regulation of metabolic pathways is often a response to changes in concentrations of key metabolites in the cell. For example, a cell may take stock of its levels of intermediate metabolites and tune the glycolytic pathway and the synthesis of glucose accordingly. In some instances, the products of a metabolic pathway actually serve as inhibitors of their own synthesis, in a process known as feedback inhibition. For example, the first intermediate in glycolysis, glucose-6-phosphate, inhibits the very enzyme that produces it, hexokinase.

References

- What-is-a-cell-14023083: nature.com, Retrieved 12 June 2018

- Cytoplasm: biologydictionary.net, Retrieved 24 July 2018

- Cell-membrane-373364: thoughtco.com, Retrieved 19 May 2018

- Cell-junctions, boundless-ap: lumenlearning.com, Retrieved 23 March 2018

- What-is-cellular-metabolism, natural-health: globalhealingcenter.com, Retrieved 31 March 2018

3

Endocrine System

The endocrine system is an organ system in the human body that produces and secretes hormones for the regulation of the activity of cells and organs. This chapter has been carefully written to provide an easy understanding of the varied aspects of the endocrine system, the endocrine glands that make up the endocrine system and the mechanism of endocrine signaling.

The endocrine system is made up of glands that produce and secrete hormones, chemical substances produced in the body that regulate the activity of cells or organs. These hormones regulate the body's growth, metabolism (the physical and chemical processes of the body), and sexual development and function. The hormones are released into the bloodstream and may affect one or several organs throughout the body.

Hormones are chemical messengers created by the body. They transfer information from one set of cells to another to coordinate the functions of different parts of the body.

The major glands of the endocrine system are the hypothalamus, pituitary, thyroid, parathyroids, adrenals, pineal body, and the reproductive organs (ovaries and testes). The pancreas is also a part of this system; it has a role in hormone production as well as in digestion.

The endocrine system is regulated by feedback in much the same way that a thermostat regulates the temperature in a room. For the hormones that are regulated by the pituitary gland, a signal is sent from the hypothalamus to the pituitary gland in the form of a "releasing hormone," which stimulates the pituitary to secrete a "stimulating hormone" into the circulation. The stimulating hormone then signals the target gland to secrete its hormone. As the level of this hormone rises in the circulation, the hypothalamus and the pituitary gland shut down secretion of the releasing hormone and the stimulating hormone, which in turn slows the secretion by the target gland. This system results in stable blood concentrations of the hormones that are regulated by the pituitary gland.

Diseases of the Endocrine System

Hormone levels that are too high or too low indicate a problem with the endocrine system. Hormone diseases also occur if your body does not respond to hormones in the appropriate ways. Stress, infection and changes in the blood's fluid and electrolyte balance can also influence hormone levels, according to the National Institutes of Health.

The most common endocrine disease in the United States is diabetes, a condition in which the body does not properly process glucose, a simple sugar. This is due to the lack of insulin or, if the body is producing insulin, because the body is not working effectively, according to Dr. Jennifer Loh, chief of the department of endocrinology for Kaiser Permanente in Hawaii.

Diabetes can be linked to obesity, diet and family history, according to Dr. Alyson Myers of North Shore-LIJ Health System. "To diagnose diabetes, we do an oral glucose tolerance test with fasting."

It is also important to understand the patient's health history as well as the family history, Myers noted. Infections and medications such as blood thinners can also cause adrenal deficiencies.

Diabetes is treated with pills or insulin injections. Managing other endocrine disorders typically involves stabilizing hormone levels with medication or, if a tumor is causing an overproduction of a hormone, by removing the tumor. Treating endocrine disorders takes a very careful and personalized approach, Myers said, as adjusting the levels of one hormone can impact the balance of other hormones.

Hormone imbalances can have a significant impact on the reproductive system, particularly in women, Loh explained.

Another disorder, hypothyroidism, a parathyroid disease, occurs when the thyroid gland does not produce enough thyroid hormone to meet the body's needs. Loh noted that insufficient thyroid hormone can cause many of the body's functions to slow or shut down completely. It has an easy treatment, though. "Parathyroid disease is a curable cause of kidney stones," said Dr. Melanie Goldfarb, an endocrine surgeon and director of the Endocrine Tumor Program at Providence Saint John's Health Center in Santa Monica, California, and an assistant professor of surgery at the John Wayne Cancer Institute in Santa Monica. The damaged part of the gland is removed surgically.

Thyroid cancer begins in the thyroid gland and starts when the cells in the thyroid begin to change, grow uncontrollably and eventually form a tumor, according to Loh. Tumors — both benign and cancerous — can also disrupt the functions of the endocrine system, Myers explained. Between the years of 1975 and 2013, the cases of thyroid cancer diagnosed yearly have more than tripled, according to a 2017 study. "While over diagnosis may be an important component to this observed epidemic, it clearly does not explain the whole story," said Dr. Julie Sosa, one of the authors of the new study and the chief of endocrine surgery at Duke University in North Carolina. The American Cancer Society predicts that there will be about 53,990 new cases of thyroid cancer in 2018 and around 2,060 deaths from thyroid cancer.

Hypoglycemia, also called low blood glucose or low blood sugar, occurs when blood glucose drops below normal levels. This typically happens as a result of treatment for diabetes when too much insulin is taken. While Loh noted that the condition can occur in people not undergoing treatment for diabetes, such an occurrence is fairly rare.

Endocrine Gland

Endocrine gland is a gland that secretes a substance (a hormone) into the bloodstream. The endocrine glands are "glands of internal secretion." They include the hypothalamus, pituitary gland, pineal gland, thyroid, parathyroid glands, heart (which makes atrial-natriuretic peptide), the stomach and intestines, islets of Langerhans in the pancreas, the adrenal glands, the kidney (which makes renin, erythropoietin, and calcitriol), fat cells (which make leptin). the testes, the ovarian follicle (estrogens) and the corpus luteum in the ovary).

Function of Endocrine Glands

The endocrine system derives its power from coordinating the interactions that take place between the hormones that are released by this network of glands. Endocrine glands themselves will inherently be able to make, secrete, and store hormones for future use. This ability to store hormones for later release is useful for modulating a response to a certain stimulus. Depending on our developmental needs at whichever stage in life we are in, our endocrine system will ensure that a proper hormonal balance is in place so that we release more or less of certain hormone based on these needs. Many factors can compromise this balance, however, resulting in endocrine disease.

One such instance is when too much or too little hormone is released from a given endocrine gland. Another problematic scenario is if an afflicted patient's blood supply is not strong enough to carry the hormones the distance they need to be carried to reach their target organs. Therefore, a vital process mediated by the endocrine system is compromised, if not many. Furthermore, once the hormones reach their target site, the tissue must have an adequate number of hormone receptors to maintain this intricate balance. The targets that receive the hormone must also be able to respond as they should to the signal. For instance, when TSH made by the pituitary gland travels through the blood to get to the thyroid, the thyroid must be able to respond by making enough thyroid hormone.

List of Endocrine Glands

Among the most important endocrine glands in the human body is the hypothalamus. In spite of its small size, this part of the brain releases crucial chemicals that influence the body's internal homeostasis as well as the pituitary gland. Its hormones include oxytocin and growth hormone, among many others. The pituitary gland, in turn, is another endocrine tissue that releases hormones related to growth, mental development, and sexual reproduction. Moving on the pineal gland in the brain, the pineal body will create and release various hormones, including melatonin, which regulates our sleep and waking cycles and eventual sexual maturation. The thyroid is an endocrine gland in the neck that releases thyroid hormones that help maintain our body's metabolic and energetic processes. The parathyroid gland, on the other hand, lies behind the thyroid gland and secretes chemicals that allow for normal bone development. The thymus has much more important roles in immune health during our childhood (via T cell production), as it is eventually phased out by fat in post pubescent children. The pancreas is another endocrine gland that will release insulin in the body, which importantly allows for sugar in the blood to be metabolized. Moving southward to the kidneys, the adrenal glands that lie above each will secrete adrenaline hormone during strenuous fight or flight situations. This modulation will likewise influence the way our bodies uses energy. Lastly, our sex organs are considered a major type of endocrine gland. Ovaries in women will create estrogen and progesterone derivatives that help with our sexual development and will aid in the release of eggs for future fertilization. Thus, all of these glands orchestrate large processes that keep our species alive and thriving. Hence, the evolutionary importance of having endocrine tissue.

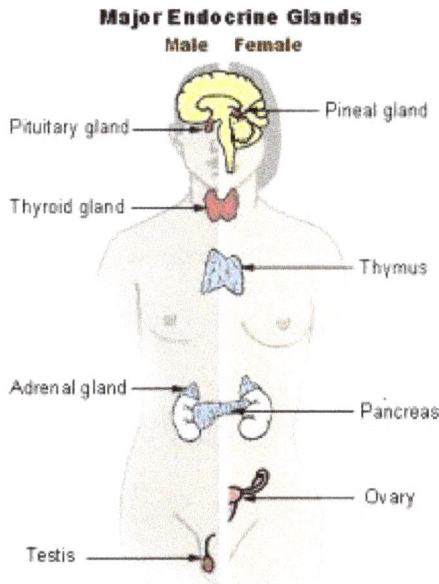

Figure: Major endocrine glands of the human body

Major Endocrine Glands

- Hypothalamus
- Thyroid
- Parathyroid

- Pituitary gland

- Adrenal glands

- Pineal gland

- Pancreas

Hypothalamus

The hypothalamus is located in the lower central part of the brain. This part of the brain is important in regulation of satiety, metabolism, and body temperature. In addition, it secretes hormones that stimulate or suppress the release of hormones in the pituitary gland. Many of these hormones are releasing hormones, which are secreted into an artery (the hypophyseal portal system) that carries them directly to the pituitary gland. In the pituitary gland, these releasing hormones signal secretion of stimulating hormones. The hypothalamus also secretes a hormone called somatostatin, which causes the pituitary gland to stop the release of growth hormone.

Pituitary Gland

Pituitary gland is ductless gland of the endocrine system that secretes hormones directly into the bloodstream. The term hypophysis (from the Greek for "lying under")—another name for the pituitary—refers to the gland's position on the underside of the brain. The pituitary gland is called the "master gland" because its hormones regulate other important endocrine glands—including the adrenal, thyroid, and reproductive glands (e.g., ovaries and testes)—and in some cases have direct regulatory effects in major tissues, such as those of the musculoskeletal system.

Anatomy of the Pituitary Gland

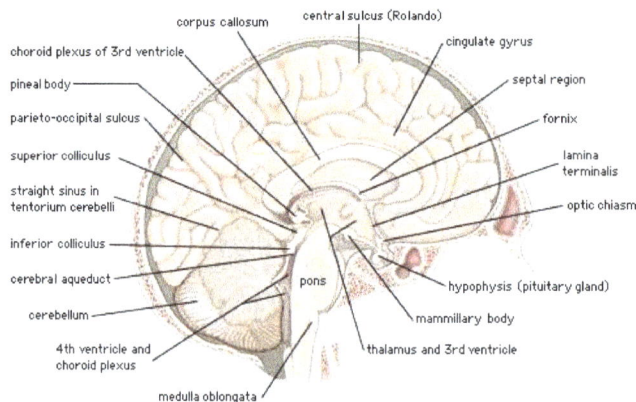

Medial view of the left hemisphere of the human brain.

The pituitary gland lies at the middle of the base of the skull and is housed within a bony structure called the sella turcica, which is behind the nose and immediately beneath the hypothalamus. The pituitary gland is attached to the hypothalamus by a stalk composed of neuronal axons and the so-called hypophyseal-portal veins. Its weight in normal adult humans ranges from about 500 to 900 mg (0.02 to 0.03 ounce).

In most species the pituitary gland is divided into three lobes: the anterior lobe, the intermediate lobe, and the posterior lobe (also called the neurohypophysis or pars nervosa). In humans, the intermediate lobe does not exist as a distinct anatomic structure but rather remains only as cells dispersed within the anterior lobe. Nonetheless, the anterior and posterior lobes of the pituitary are functionally, anatomically, and embryologically distinct. Whereas the anterior pituitary contains abundant hormone-secreting epithelial cells, the posterior pituitary is composed largely of unmyelinated (lacking a sheath of fatty insulation) secretory neurons.

Thyroid Gland

Thyroid gland is a gland that makes and stores hormones that help regulate the heart rate, blood pressure, body temperature, and the rate at which food is converted into energy. Thyroid hormones are essential for the function of every cell in the body. They help regulate growth and the rate of chemical reactions (metabolism) in the body. Thyroid hormones also help children grow and develop.

The thyroid gland is located in the lower part of the neck, below the Adam's apple, wrapped around the trachea (windpipe). It has the shape of a butterfly: two wings (lobes) attached to one another by a middle part called the isthmus.

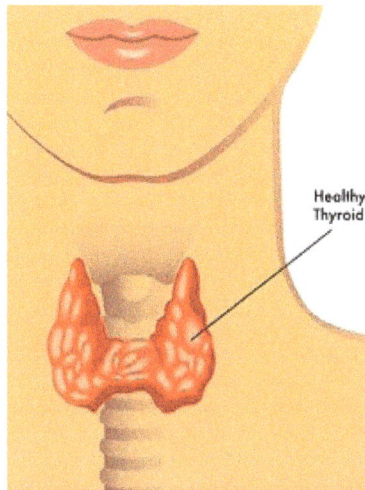

Healthy
Thyroid

The thyroid's hormones regulate vital body functions, including:

- Breathing

- Heart rate

- Central and peripheral nervous systems

- Body weight

- Muscle strength

- Menstrual cycle

- Body temperature

- Cholesterol levels

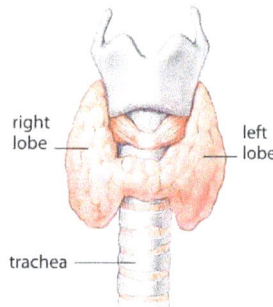

The thyroid gland is about 2-inches long and lies in front of your throat below the prominence of thyroid cartilage sometimes called the Adam's apple. The thyroid has two sides called lobes that lie on either side of your windpipe, and is usually connected by a strip of thyroid tissue known as an isthmus. Some people do not have an isthmus, and instead have two separate thyroid lobes.

Working of Thyroid Gland

The thyroid is part of the endocrine system, which is made up of glands that produce, store, and release hormones into the bloodstream so the hormones can reach the body's cells. The thyroid gland uses iodine from the foods you eat to make two main hormones:

- Triiodothyronine (T3)

- Thyroxine (T4)

It is important that T3 and T4 levels are neither too high nor too low. Two glands in the brain—the hypothalamus and the pituitary communicate to maintain T3 and T4 balance.

The hypothalamus produces TSH Releasing Hormone (TRH) that signals the pituitary to tell the thyroid gland to produce more or less of T3 and T4 by either increasing or decreasing the release of a hormone called thyroid stimulating hormone (TSH).

- When T3 and T4 levels are low in the blood, the pituitary gland releases more TSH to tell the thyroid gland to produce more thyroid hormones.

- If T3 and T4 levels are high, the pituitary gland releases less TSH to the thyroid gland to slow production of these hormones.

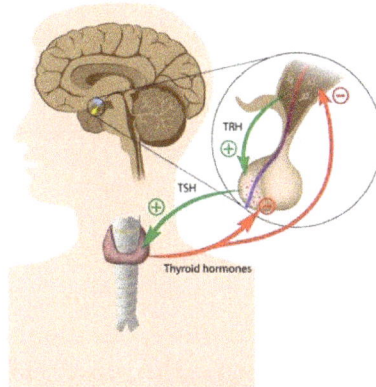

Need of a Thyroid Gland

T3 and T4 travel in the bloodstream to reach almost every cell in the body. The hormones regulate the speed with which the cells/metabolism work. For example, T3 and T4 regulate your heart rate and how fast your intestines process food. So if T3 and T4 levels are low, your heart rate may be slower than normal, and you may have constipation/weight gain. If T3 and T4 levels are high, you may have a rapid heart rate and diarrhea/weight loss.

Listed below are other symptoms of too much T3 and T4 in your body (hyperthyroidism):

- Anxiety
- Irritability or moodiness
- Nervousness, hyperactivity
- Sweating or sensitivity to high temperatures
- Hand trembling (shaking)
- Hair loss
- Missed or light menstrual periods

The following are other symptoms that may indicate too little T3 and T4 in your body (hypothyroidism):

- Trouble sleeping
- Tiredness and fatigue
- Difficulty concentrating
- Dry skin and hair
- Depression

- Sensitivity to cold temperature

- Frequent, heavy periods

- Joint and muscle pain

Parathyroid Gland

A gland that regulates calcium, located behind the thyroid gland in the neck. The parathyroid gland secretes a hormone called parathormone (or parathyrin) that is critical to calcium and phosphorus metabolism. Although the number of parathyroid glands can vary, most people have four, one above the other on each side. They are plastered against the back of the thyroid and therefore at risk for being accidentally removed during thyroidectomy.

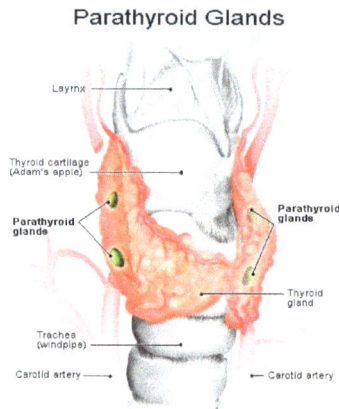

Parathyroid Glands

Laynx

Thyroid cartilage
(Adam's apple)

Parathyroid
glands

Parathyroid
glands

Thyroid
gland

Trachea
(windpipe)

Carotid artery

Carotid artery

Work of Parathyroid Glands

Parathyroid glands control the calcium levels in our blood, in our bones, and throughout our body. Parathyroid glands regulate the calcium by producing a hormone called Parathyroid Hormone (PTH). Calcium is the most important element in our bodies (we use it to control many organ systems), so calcium is regulated more carefully than any other element. In fact, calcium is the only element with it's own regulatory system -- the parathyroid glands.

Locations of Parathyroid Glands

Parathyroid glands (we all have 4 of them) are normally the size of a grain of rice. Occasionally they can be as large as a pea and still be normal. The four parathyroids are located behind the thyroid and are shown in this picture as the mustard yellow glands behind the pink thyroid gland. Normal parathyroid glands are the color of spicy yellow mustard. The light blue tube running up the center of the picture is the trachea (wind pipe). The voice box is the pink structure at the top of the picture sitting on top of the trachea. The carotid arteries are shown on both sides of the thyroid running from the heart up to the brain. Remember, the parathyroids are behind the thyroid. If you have parathyroid disease, you very likely have 3 normal parathyroid glands the size of a grain of rice and one parathyroid tumor that is as big as an olive, grape, or even a walnut. If you have parathyroid disease (hyperparathyroidism) you will need an operation to remove the one parathyroid gland which has become a tumor. Although they are neighbors and both are part of the endocrine system, the thyroid and parathyroid glands are otherwise unrelated--they do not have the same function.

Adrenal Gland

Adrenal gland is either of two small triangular endocrine glands one of which is located above each kidney. In humans each adrenal gland weighs about 5 grams (0.18 ounce) and measures about 30 mm (1.2 inches) wide, 50 mm (2 inches) long, and 10 mm (0.4 inch) thick. Each gland consists of two parts: an inner medulla, which produces epinephrine and norepinephrine (adrenaline and noradrenaline), and an outer cortex, which produces steroid hormones. The two parts differ in embryological origin, structure, and function. The adrenal glands vary in size, shape, and nerve supply in other animal species. In some vertebrates the cells of the two parts are interspersed to varying degrees.

Adrenal Cortex Hormones

The adrenal cortex produces two main groups of corticosteroid hormones—glucocorticoids and mineralcorticoids. The release of glucocorticoids is triggered by the hypothalamus and pituitary gland. Mineralcorticoids are mediated by signals triggered by the kidney.

When the hypothalamus produces corticotrophin-releasing hormone (CRH), it stimulates the pituitary gland to release adrenal corticotrophic hormone (ACTH). These hormones, in turn, alert the adrenal glands to produce corticosteroid hormones.

Glucocorticoids released by the adrenal cortex include:

- Hydrocortisone: Commonly known as cortisol, it regulates how the body converts fats, proteins, and carbohydrates to energy. It also helps regulate blood pressure and cardiovascular function.

- Corticosterone: This hormone works with hydrocortisone to regulate immune response and suppress inflammatory reactions.

The principle mineralcorticoid is aldosterone, which maintains the right balance of salt and water while helping control blood pressure.

There is a third class of hormone released by the adrenal cortex, known as sex steroids or sex hormones. The adrenal cortex releases small amounts of male and female sex hormones. However, their impact is usually overshadowed by the greater amounts of hormones (such as estrogen and testosterone) released by the ovaries or testes.

Adrenal Medulla Hormones

Unlike the adrenal cortex, the adrenal medulla does not perform any vital functions. That is, you don't need it to live. But that hardly means the adrenal medulla is useless. The hormones of the adrenal medulla are released after the sympathetic nervous system is stimulated, which occurs when you're stressed. As such, the adrenal medulla helps you deal with physical and emotional stress.

You may be familiar with the fight-or-flight response—a process initiated by the sympathetic nervous system when your body encounters a threatening (stressful) situation. The hormones of the adrenal medulla contribute to this response.

Hormones secreted by the adrenal medulla are:

- Epinephrine: Most people know epinephrine by its other name—adrenaline. This hormone rapidly responds to stress by increasing your heart rate and rushing blood to the muscles and brain. It also spikes your blood sugar level by helping convert glycogen to glucose in the liver. (Glycogen is the liver's storage form of glucose.)

- Norepinephrine: Also known as noradrenaline, this hormone works with epinephrine in responding to stress. However, it can cause vasoconstriction (the narrowing of blood vessels). This results in high blood pressure.

Pineal Gland

Pineal gland is also known as Conarium, Epiphysis cerebri, pineal organ or pineal body or third eye. It is an endocrine gland found in vertebrates that is the source of melatonin, a hormone derived from tryptophan that plays a central role in the regulation of circadian rhythm (the roughly 24-hour cycle of biological activities associated with natural periods of light and darkness).

The pineal gland has long been an enigmatic structure. Even in the early 21st century, when sophisticated molecular techniques were available for biological study, fundamental features of the gland—including the extent of the effects of its principal hormone, melatonin—remained incompletely understood.

Anatomy of the Pineal Gland

The pineal gland develops from the roof of the diencephalon, a section of the brain, and is located behind the third cerebral ventricle in the brain midline (between the two cerebral hemispheres). Its name is derived from its shape, which is similar to that of a pinecone (Latin pinea). In adult humans it is about 0.8 cm (0.3 inch) long and weighs approximately 0.1 gram (0.004 ounce).

The pineal gland has a rich supply of adrenergic nerves (neurons sensitive to the adrenal hormone epinephrine) that greatly influence its function. Microscopically, the gland is composed of pinealocytes (rather typical endocrine cells except for extensions that mingle with those of adjacent cells) and supporting cells that are similar to the astrocytes of the brain. In adults, small deposits of calcium often make the pineal body visible on X-rays. (The pineal gland eventually becomes more or less calcified in most people.)

In some lower vertebrates, the gland has a well-developed eyelike structure. In others, though not organized as an eye, it functions as a light receptor.

Pineal Hormones

Both melatonin and its precursor, serotonin, which are derived chemically from the alkaloid substance tryptamine, are synthesized in the pineal gland. Along with other brain sites, the pineal gland may also produce neurosteroids. Dimethyltryptamine (DMT), a hallucinogenic compound present in the Amazonian botanical drink ayahuasca, is chemically similar to melatonin and serotonin and is considered to be a trace substance in human blood and urine. Although alleged to be produced by the pineal gland, DMT has not been consistently detected in human pineal microdialysates (purified pineal extracts), and proof of its regulated biosynthesis in the mammalian pineal gland is lacking. Thus, though the conclusion by 17th-century French philosopher René Descartes that the pineal gland is the seat of the soul has endured as a historical curiosity, there is no evidence to support the notion that secretions from the pineal have a major role in cognition.

In addition to the pineal gland, melatonin is also synthesized in the vertebrate retina, where it transduces information about environmental light through local receptors designated MT1 and MT2, and in certain other tissues, such as the gastrointestinal tract and the skin. In the generally rate-limiting step of melatonin biosynthesis, an enzyme called serotonin N-acetyltransferase (AANAT) catalyzes the conversion of serotonin to N-acetylserotonin. Subsequently that compound is catalyzed to melatonin by acetylserotonin O-methyltransferase (ASMT). The rise in circulating melatonin concentrations that occurs and is maintained after sundown and with darkness coincides with the activation of AANAT during dark periods. Melatonin concentrations also are higher in the cerebrospinal fluid (CSF) of the third ventricle of the brain than in the CSF of the fourth ventricle or in the blood. That suggests that melatonin is also secreted directly into the CSF, where it may have direct and perhaps more-sustained effects on target areas of the central nervous system.

In some species pineal cells are photosensitive. In humans and higher mammals a "photoendocrine system"—made up of the retina, the suprachiasmatic nucleus of the hypothalamus, and noradrenergic sympathetic fibres (neurons responsive to the neurotransmitter norepinephrine) terminating in the pineal—provides light and circadian information that regulates pineal melatonin secretion. In contrast to many other endocrine hormones, human melatonin concentrations are highly variable, and serum melatonin levels decline markedly during childhood, as there is little or no growth of the pineal gland after about one year of age.

Pancreas

The pancreas is a long flattened gland located deep in the belly (abdomen). Because the pancreas isn't seen or felt in our day to day lives, most people don't know as much about the pancreas as they do about other parts of their bodies. The pancreas is, however, a vital part of the digestive system and a critical controller of blood sugar levels.

The pancreas is located deep in the abdomen. Part of the pancreas is sandwiched between the stomach and the spine. The other part is nestled in the curve of the duodenum (first part of the small intestine). To visualize the position of the pancreas, try this: touch your right thumb and right "pinkie" fingers together, keeping the other three fingers together and straight. Then, place your hand in the center of your belly just below your lower ribs with your fingers pointing to your left. Your hand will be the approximate shape and at the approximate level of your pancreas.

Because of the deep location of the pancreas, tumors of the pancreas are rarely palpable (able to be felt by pressing on the abdomen). This explains why most symptoms of pancreatic cancer do not appear until the tumor has grown large enough to interfere with the function of the pancreas or other nearby organs such as the stomach, duodenum, liver, or gallbladder.

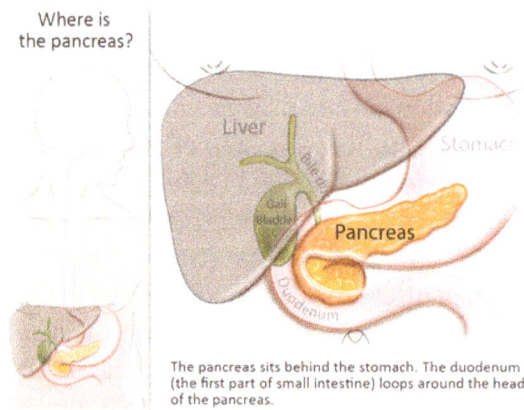

Where is the pancreas?

Liver

Stomach

Bile duct

Gall Bladder

Pancreas

Duodenum

The pancreas sits behind the stomach. The duodenum (the first part of small intestine) loops around the head of the pancreas.

Functions of the Pancreas

A healthy pancreas produces the correct chemicals in the proper quantities, at the right times, to digest the foods we eat.

- Exocrine Function

The pancreas contains exocrine glands that produce enzymes important to digestion. These enzymes include trypsin and chymotrypsin to digest proteins; amylase for the digestion of carbohydrates; and lipase to break down fats. When food enters the stomach, these pancreatic juices are released into a system of ducts that culminate in the main pancreatic duct. The pancreatic duct joins the common bile duct to form the ampulla of Vater which is located at the first portion of the small intestine, called the duodenum. The common bile duct originates in the liver and the gallbladder and produces another important digestive juice called bile. The pancreatic juices and bile that are released into the duodenum help the body to digest fats, carbohydrates, and proteins.

- Endocrine Function

The endocrine component of the pancreas consists of islet cells (islets of Langerhans) that create and release important hormones directly into the bloodstream. Two of the main pancreatic hormones are insulin, which acts to lower blood sugar, and glucagon, which acts to raise blood sugar. Maintaining proper blood sugar levels is crucial to the functioning of key organs including the brain, liver, and kidneys.

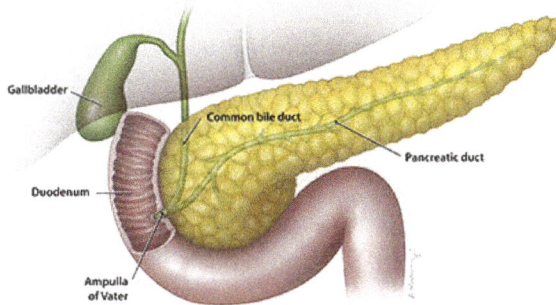

The pancreas, gallbladder and duodenum

Endocrine Signaling

When cells need to transmit signals over long distances, they often use the circulatory system as a distribution network for the messages they send. In long-distance endocrine signaling, signals are produced by specialized cells and released into the bloodstream, which carries them to target cells in distant parts of the body. Signals that are produced in one part of the body and travel through the circulation to reach far-away targets are known as hormones.

In humans, endocrine glands that release hormones include the thyroid, the hypothalamus, and the pituitary, as well as the gonads (testes and ovaries) and the pancreas. Each endocrine gland releases one or more types of hormones, many of which are master regulators of development and physiology.

For example, the pituitary releases growth hormone (GH), which promotes growth, particularly of the skeleton and cartilage. Like most hormones, GH affects many different types of cells throughout the body. However, cartilage cells provide one example of how GH functions: it binds to receptors on the surface of these cells and encourages them to divide.

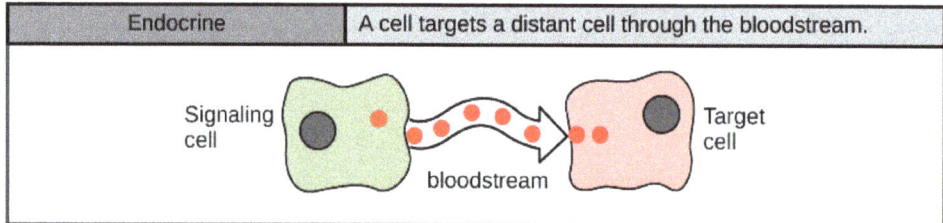

Figure: Signaling molecules and cellular receptors

References

- Anatomy-of-the-endocrine-system: emedicinehealth.com, Retrieved 04 July 2018

- Endocrine-system-26496: livescience.com, Retrieved 23 April 2018

- Thyroid-gland-controls-bodys-metabolism-how-it-works-symptoms-hyperthyroi: endocrineweb.com, Retrieved 16 July 2018

- Adrenal-gland, science: britannica.com, Retrieved 11 June 2018

- Pancreas-and-its-functions: columbiasurgery.org, Retrieved 24 May 2018

- Introduction-to-cell-signaling, mechanisms-of-cell-signaling, cell-signaling: khanacademy.org, Retrieved 18 June 2018

4

Muscular System

The muscular system of the human body is a major organ system that controls the movement of the body, circulates blood and maintains posture. This chapter closely examines the central constituents of the muscular system, such as smooth muscle, cardiac muscle, skeletal muscle, etc.

The muscular system is responsible for the movement of the human body. Attached to the bones of the skeletal system are about 700 named muscles that make up roughly half of a person's body weight. Each of these muscles is a discrete organ constructed of skeletal muscle tissue, blood vessels, tendons, and nerves. Muscle tissue is also found inside of the heart, digestive organs, and blood vessels. In these organs, muscles serve to move substances throughout the body.

Movement

The most obvious function of the muscular system is movement. Organisms have adopted a variety of methods to use the contractile function of the muscular system to move themselves through the environment. The most basic movements of fish include contracting muscles on opposite sides of the body in succession to propel themselves through the water. In organisms with limbs, tendons and other connective tissues are used to secure muscles to the joints and skeleton, which may be internal like human skeletons, or external like the exoskeleton of crabs. The nervous system coordinates the contraction of the muscular system to synchronize the movement of the limbs. Animals like the cheetah, swordfish, and bat have obtained speeds above 60 miles per hour through the power of their muscles alone.

Circulation

The second and less obvious function of the muscular system is to assist with circulation. Visceral and cardiac muscle tissues surround the blood vessels and lymph vessels that carry crucial nutrients to the cells of the body. Cardiac muscle makes up the heart, and supplies the main force for blood traveling through the body. Large arteries and veins have associated muscles which can contract or relax to control the pressure of the blood. The actions of large skeletal muscles also help pump the blood and lymph through the body. While you exercise and contract large and small muscles, they push vessels aside, which works like a pump to move fluids around your body.

Digestion

Much like its ability to move fluids through vessels in the circulatory system, the muscular system also aids in moving food through the digestive system. Most digestive organs are surrounded by visceral, or smooth muscle tissue. Although the tissue cannot be voluntarily contracted like skeletal muscles, it is controlled subconsciously. When food needs to be moved through the gut, the muscles contract in a synchronized fashion in a wave through the digestive system.

Muscle Anatomy

A muscle fiber is simply a muscle cell, or the building blocks of muscles. See, just like the rest of your body is made up of lots and lots of individual cells, your muscles are also made up of cells.

Multiple fibers join together to make up our next level of organization, the fascicle. A fascicle is a group, or bundle, of muscle fibers. And, just like many muscle fibers join together to form a fascicle, many fascicles join together to make up a muscle.

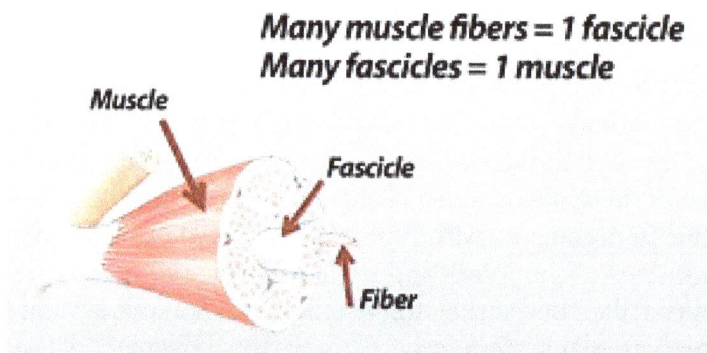

Muscles are composed of fibers and fascicles.

Muscles are made up of groups of fascicles. When stimulated, they contract to produce motion. Without muscles, your body wouldn't be able to function. No more running, talking, walking and, well, you get the picture.

The organization of the muscle allows all the fibers, and thus the fascicles, to contract and relax as a group. Contraction is stimulated by nerve impulses and triggers the movement of the muscle, while relaxation occurs when the impulse is removed and the muscle relaxes back to its natural state. This pattern of contraction and relaxation is responsible for all the movements in your body.

The part of your body that moves in response to a muscle contraction depends on the location and origin point of the muscles themselves. To simplify things, we are going to focus on the skeletal muscles of the body. Most skeletal muscles are attached to bone, cartilage or connective tissue, which limits or directs their movement. For example, a muscle attached to the arm bone will only move the arm bone when stimulated. It cannot move the leg bone; therefore, its movement is determined by its points of attachment.

Smooth Muscle

Smooth muscle (so-named because the cells do not have striations) is present in the walls of hollow organs like the urinary bladder, uterus, stomach, intestines, and in the walls of passageways, such as the arteries and veins of the circulatory system, and the tracts of the respiratory, urinary, and reproductive systems. Smooth muscle is also present in the eyes, where it functions to change the size of the iris and alter the shape of the lens; and in the skin where it causes hair to stand erect in response to cold temperature or fear.

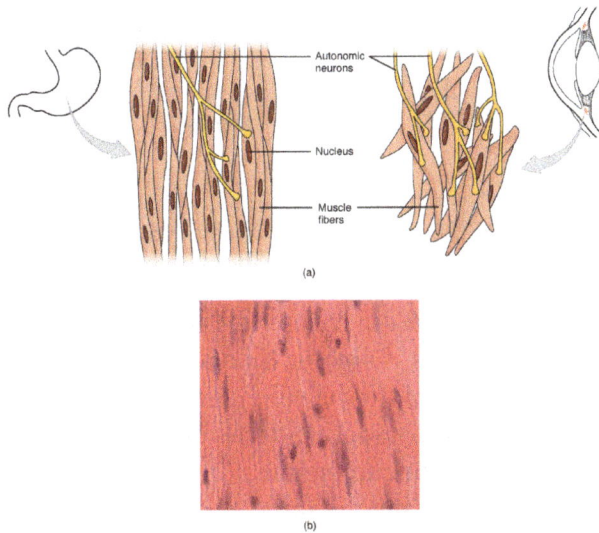

Figure: Smooth Muscle Tissue

Smooth muscle tissue is found around organs in the digestive, respiratory, reproductive tracts and the iris of the eye. LM × 1600.

Smooth muscle fibers are spindle-shaped (wide in the middle and tapered at both ends, somewhat like a football) and have a single nucleus; they range from about 30 to 200 μm (thousands of times shorter than skeletal muscle fibers), and they produce their own connective tissue, endomysium. Although they do not have striations and sarcomeres, smooth muscle fibers do have actin and myosin contractile proteins, and thick and thin filaments. These thin filaments are anchored by dense bodies. A dense body is analogous to the Z-discs of skeletal and cardiac muscle fibers and is fastened to the sarcolemma. Calcium ions are supplied by the SR in the fibers and by sequestration from the extracellular fluid through membrane indentations called calveoli.

Because smooth muscle cells do not contain troponin, cross-bridge formation is not regulated by the troponin-tropomyosin complex but instead by the regulatory protein calmodulin. In a smooth muscle fiber, external Ca^{++} ions passing through opened calcium channels in the sarcolemma, and additional Ca^{++} released from SR, bind to calmodulin. The Ca^{++}-calmodulin complex then activates an enzyme called myosin (light chain) kinase, which, in turn, activates the myosin heads by phosphorylating them (converting ATP to ADP and P_i, with the P_i attaching to the head). The heads can then attach to actin-binding sites and pull on the thin filaments. The thin filaments also are anchored to the dense bodies; the structures invested in the inner membrane of the sarcolemma (at adherens junctions) that also have cord-like intermediate filaments attached to them. When the thin filaments slide past the thick filaments, they pull on the dense bodies, structures tethered to the sarcolemma, which then pull on the intermediate filaments networks throughout the sarcoplasm. This arrangement causes the entire muscle fiber to contract in a manner whereby the ends are pulled toward the center, causing the midsection to bulge in a corkscrew motion.

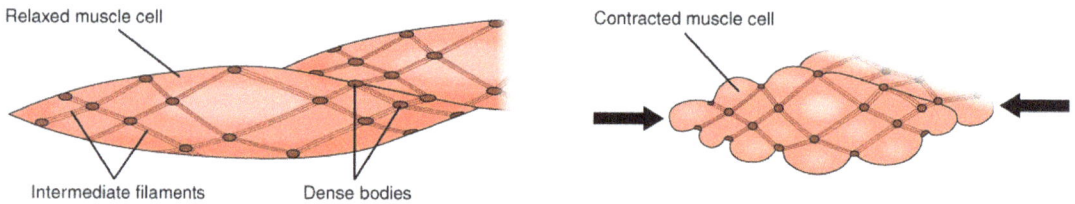

Figure: Muscle Contraction. The dense bodies and intermediate filaments are networked through the sarcoplasm, which cause the muscle fiber to contract.

Although smooth muscle contraction relies on the presence of Ca^{++} ions, smooth muscle fibers have a much smaller diameter than skeletal muscle cells. T-tubules are not required to reach the interior of the cell and therefore not necessary to transmit an action potential deep into the fiber. Smooth muscle fibers have a limited calcium-storing SR but have calcium channels in the sarcolemma (similar to cardiac muscle fibers) that open during the action potential along the sarcolemma. The influx of extracellular Ca^{++} ions, which diffuse into the sarcoplasm to reach the calmodulin, accounts for most of the Ca^{++} that triggers contraction of a smooth muscle cell.

Muscle contraction continues until ATP-dependent calcium pumps actively transport Ca^{++} ions back into the SR and out of the cell. However, a low concentration of calcium remains in the sarcoplasm to maintain muscle tone. This remaining calcium keeps the muscle slightly contracted, which is important in certain tracts and around blood vessels.

Because most smooth muscles must function for long periods without rest, their power output is relatively low, but contractions can continue without using large amounts of energy. Some smooth muscle can also maintain contractions even as Ca^{++} is removed and myosin kinase is inactivated/dephosphorylated. This can happen as a subset of cross-bridges between myosin heads and actin, called latch-bridges, keep the thick and thin filaments linked together for a prolonged period, and without the need for ATP. This allows for the maintaining of muscle "tone" in smooth muscle that lines arterioles and other visceral organs with very little energy expenditure.

Smooth muscle is not under voluntary control; thus, it is called involuntary muscle. The triggers for smooth muscle contraction include hormones, neural stimulation by the ANS, and local factors. In certain locations, such as the walls of visceral organs, stretching the muscle can trigger its contraction (the stretch-relaxation response).

Axons of neurons in the ANS do not form the highly organized NMJs with smooth muscle, as seen between motor neurons and skeletal muscle fibers. Instead, there is a series of neurotransmitter-filled bulges called varicosities as an axon courses through smooth muscle, loosely forming motor units. A varicosity releases neurotransmitters into the synaptic cleft. Also, visceral muscle in the walls of the hollow organs (except the heart) contains pacesetter cells. A pacesetter cell can spontaneously trigger action potentials and contractions in the muscle.

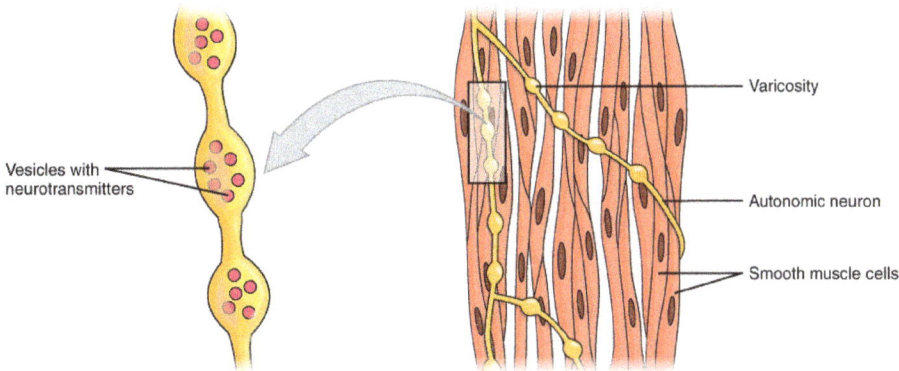

Figure: Motor Units. A series of axon-like swelling, called varicosities or "boutons,"
from autonomic neurons form motor units through the smooth muscle.

Smooth muscle is organized in two ways: as single-unit smooth muscle, which is much more common; and as multiunit smooth muscle. The two types have different locations in the body and have different characteristics. Single-unit muscle has its muscle fibers joined by gap junctions so that the muscle contracts as a single unit. This type of smooth muscle is found in the walls of all visceral organs except the heart (which has cardiac muscle in its walls), and so it is commonly called visceral muscle. Because the muscle fibers are not constrained by the organization and stretchability limits of sarcomeres, visceral smooth muscle has a stress-relaxation response. This means that as the muscle of a hollow organ is stretched when it fills, the mechanical stress of the stretching will trigger contraction, but this is immediately followed by relaxation so that the organ does not empty its contents prematurely. This is important for hollow organs, such as the stomach or urinary bladder, which continuously expand as they fill. The smooth muscle around these organs also can maintain a muscle tone when the organ empties and shrinks, a feature that prevents "flabbiness" in the empty organ. In general, visceral smooth muscle produces slow, steady contractions that allow sub-stances, such as food in the digestive tract, to move through the body.

Multiunit smooth muscle cells rarely possess gap junctions, and thus are not electri-cally coupled. As a result, contraction does not spread from one cell to the next, but is instead confined to the cell that was originally stimulated. Stimuli for multiunit smooth muscles come from autonomic nerves or hormones but not from stretching. This type of tissue is found around large blood vessels, in the respiratory airways, and in the eyes.

Hyperplasia in Smooth Muscle

Similar to skeletal and cardiac muscle cells, smooth muscle can undergo hypertrophy to increase in size. Unlike other muscle, smooth muscle can also divide to produce more cells, a process called hyperplasia. This can most evidently be observed in the uterus at puberty, which responds to increased estrogen levels by producing more uter-ine smooth muscle fibers, and greatly increases the size of the myometrium.

Function of Smooth Muscle

Like all muscle tissue, the function of smooth muscle is to contract. The image above shows how the actin and myosin fibers shorten, effectively shrinking the cell. However, there are some important differences in how the smooth muscle contracts, compared to other types of muscle. In skeletal muscle, a signal from the somatic nervous system traverses to the muscle, where it stimulates organelles in the muscle cell to release calcium. The calcium causes a protein to detach from actin, and myosin quickly binds to the opening on actin. Since there was always available ATP, the myosin uses it to quickly contract the cell.

The same is not true in smooth muscle tissue. In smooth muscle, the contraction is not controlled voluntarily by the somatic nervous system, but by signals from the au-tonomous nervous system, such as nerve impulses, hormones, and other chemicals released by specialized organs. Smooth muscle is specialized to contract persistently, unlike skeletal muscle which much contract and release quickly. Instead of a calcium trigger which sets off a contraction reaction, smooth muscle has more of a throttle, like in a car.

A nerve impulse or outside stimulus reaches the cell, which tells it to release calcium. Smooth muscle cells do not have a special protein on actin which prevents myosin from binding. Rather, actin and myosin are constantly binding. But, myosin can only hold on and crawl forward when given energy. Inside smooth muscle cells is a complex pathway which allows the level of calcium to control the amount of ATP available to myosin. Thus, when the stimulus is removed, the cells do not relax right away. Myosin contin-ues to bind to actin and crawl along the filaments until the level of calcium falls.

Smooth Muscle Location

This specialized function of contracting for long periods and hold that force is why smooth muscle has been adapted to many areas of the body. Smooth muscle lines many parts of the circulatory system, digestive system, and is even responsible for raising the hairs on your arm.

In the circulatory system, smooth muscle plays a vital role in maintaining and controlling the blood pressure and flow of oxygen throughout the body. While the majority of the pressure is applied by the heart, every vein and artery is lined with smooth muscle. These small muscles can contract to apply pressure to the system or relax to allow more blood to flow. Tests have shown that these smooth muscles are stimulated by the presence or absence of oxygen, and modify the veins to provide enough oxygen when it is low.

Smooth muscle also lines the majority of the digestive system, for similar reasons. However, the cells in the digestive system have different stimuli than those in the circulatory system. For instance, sheets of smooth muscle tissue in the gut react to you swallowing. When you swallow, tension is applied to one side of the sheet. The cells on that side contract in reaction, a wave begins to propagate itself down your digestive tract. This phenomena is known as peristalsis, and is responsible for moving food through the many twists and turns of the gut.

Smooth muscle, because of its ability to contract and hold, is used for many function in many places of the body. Besides those listed above, smooth muscle is also responsible for contracting the irises, raising the small hairs on your arm, contracting the many sphincters in your body, and even moving fluids through organs by applying pressure to them. While smooth muscle doesn't contract or release as quickly as skeletal or cardiac muscle, it is much more useful for providing consistent, elastic tension.

Cardiac Muscle

Cardiac muscle, also known as heart muscle, is the layer of muscle tissue which lies between the endocardium and epicardium. These inner and outer layers of the heart, respectively, surround the cardiac muscle tissue and separate it from the blood and other organs. Cardiac muscle is made from sheets of cardiac muscle cells. These cells, unlike skeletal muscle cells, are typically unicellular and connect to one another through special intercalated discs. These specialized cell junction and the arrangement of muscle cells enables cardiac muscle to contract quickly and repeatedly, forcing blood throughout the body.

Cardiac Muscle Structure

Cardiac muscle exists only within the heart of animals. It is a specialized form of muscle evolved to continuously and repeatedly contract, providing circulation of blood throughout the body. The heart is a relatively simple organ. Through all the twists and turns and various chambers, there are only three layers. The outer layer, known as the epicardium or visceral pericardium, surround the cardiac muscle on the exterior. This helps protect it from contact with other organs. The parietal pericardium attaches to this outer layer creates a fluid-filled layer which helps lubricate the heart. The inner layer, or endocardium, separates the muscle from the blood it is pumping within the chambers of the heart. In between these two sheets lies the cardiac muscle. Cardiac muscle is sometimes referred to as myocardium.

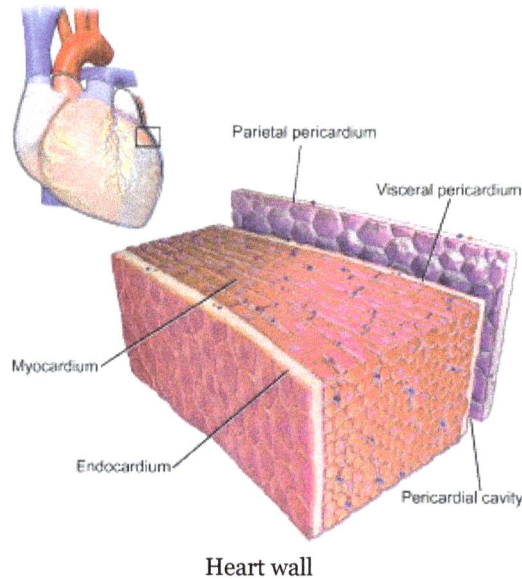

Heart wall

When we look a bit closer at cardiac muscle, we can see that it is arranged in sheets of cells, which are connected to each other in a lattice-work fashion. Where two cells meet a specialized junction called an intercalated disc locks the two cells into place. While this region looks like a dark disc under the microscope, it is actually the interlocking of hundreds of finger-like projections from each cell. These projections have small holes in them, gap junctions, which can pass the impulse to contract to connected cells. Interlaced between and around these cells are nerves and blood vessels, which carry signals and oxygen to the cardiac muscle.

At the microscopic level, cardiac muscle is organized much like skeletal muscle. Both muscle tissues are striated, meaning they show dark and light bands when viewed under a microscope. These band are created by the highly organized sarcomeres. A sarcomere is a bundle of protein fibers which respond to a signal and contract. In both skeletal and cardiac muscle, these sarcomeres are made of actin and myosin and are supported by the same proteins. Tropomyosin is a protein which wraps actin and stops myosin from binding to it. Troponin is a protein which holds tropomyosin in place until a signal to contract has been received. These proteins are the same in both skeletal and cardiac muscle.

Function of Cardiac Muscle

As in skeletal muscle, the signal to contract is an action potential. However with skeletal muscle this signal usually comes from the somatic, or voluntary, nervous system. Cardiac muscle is controlled by the autonomous nervous system. Cells in your brain and cells embedded throughout your heart act to release well-timed nervous impulses which signal your heart cells to contract in the correct pattern. While the source of the signals is different, the reception of the signal and the rest of contraction are very similar.

The action potential, or nerve impulse, on the surface of the cell stimulates a specialized organelle to release calcium ions (Ca^{2+}). This organelle is called the sarcoplasmic reticulum, and is derived from the endoplasmic reticulum found in a general cell. The Ca^{2+} ions released into the cytoplasm affect the protein troponin, causing it to release tropomyosin. Tropomyosin shifts position and myosin is allowed to attach to actin. Myosin then used the energy stored in ATP molecules to walk along the actin filaments and shorten the length of each sarcomere. When the impulse is gone, the Ca^{2+} is reabsorbed quickly into the sarcoplasmic reticulum. Troponin reattaches to tropomyosin, and the cardiac muscle cells release. This general process happens every time your heart beats.

As all the muscle cells work in unison, a force can be exerted in the chambers of the heart. The sheets of cardiac muscle are laid so they run perpendicularly to one another. This creates the effect that when the heart contracts, it does so in multiple directions. The ventricles and atria of the heart shrink from top to bottom and from side to side as these multiple layers muscle fibers contract. This produces a strong pumping and twisting force in the ventricles, forcing blood throughout the body.

Skeletal Muscle

Skeletal muscle is also called voluntary muscle. In vertebrates, it is most common of the three types of muscle in the body. Skeletal muscles are attached to bones by tendons, and they produce all the movements of body parts in relation to each other. Unlike smooth muscle and cardiac muscle, skeletal muscle is under voluntary control. Similar to cardiac muscle, however, skeletal muscle is striated; its long, thin, multinucleated fibres are crossed with a regular pattern of fine red and white lines, giving the muscle a distinctive appearance. Skeletal muscle fibres are bound together by connective tissue and communicate with nerves and blood vessels.

Structure of Skeletal Muscle

A whole skeletal muscle is considered an organ of the muscular system. Each organ or muscle consists of skeletal muscle tissue, connective tissue, nerve tissue, and blood or vascular tissue.

Skeletal muscles vary considerably in size, shape, and arrangement of fibers. They range from extremely tiny strands such as the stapedium muscle of the middle ear to large masses such as the muscles of the thigh. Some skeletal muscles are broad in shape and some narrow. In some muscles, the fibers are parallel to the long axis of the muscle; in some they converge to a narrow attachment; and in some they are oblique.

Structure of a Skeletal Muscle

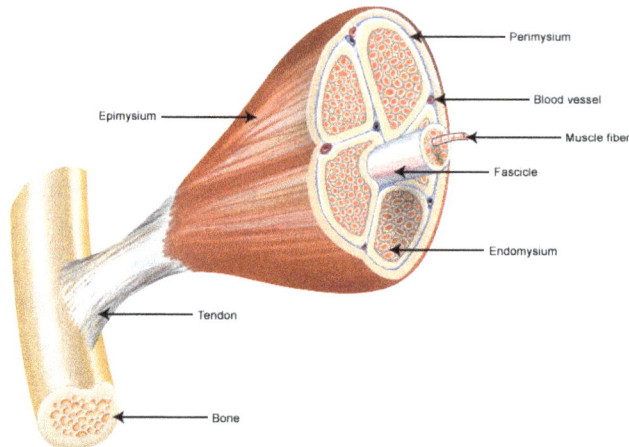

Each skeletal muscle fiber is a single cylindrical muscle cell. An individual skeletal muscle may be made up of hundreds, or even thousands, of muscle fibers bundled together and wrapped in a connective tissue covering. Each muscle is surrounded by a connective tissue sheath called the epimysium. Fascia, connective tissue outside the epimysium, surrounds and separates the muscles. Portions of the epimysium project inward to divide the muscle into compartments. Each compartment contains a bundle of muscle fibers. Each bundle of muscle fiber is called a fasciculus and is surrounded by a layer of connective tissue called the perimysium. Within the fasciculus, each individual muscle cell, called a muscle fiber, is surrounded by connective tissue called the endomysium.

Skeletal muscle cells (fibers), like other body cells, are soft and fragile. The connective tissue covering furnish support and protection for the delicate cells and allow them to withstand the forces of contraction. The coverings also provide pathways for the passage of blood vessels and nerves.

Commonly, the epimysium, perimysium, and endomysium extend beyond the fleshy part of the muscle, the belly or gaster, to form a thick ropelike tendon or a broad, flat sheet-like aponeurosis. The tendon and aponeurosis form indirect attachments from muscles to the periosteum of bones or to the connective tissue of other muscles. Typically, a muscle spans a joint and is attached to bones by tendons at both ends. One of the bones remains relatively fixed or stable while the other end moves as a result of muscle contraction.

Skeletal muscles have an abundant supply of blood vessels and nerves. This is directly related to the primary function of skeletal muscle, contraction. Before a skeletal muscle fiber can contract, it has to receive an impulse from a nerve cell. Generally, an artery and at least one vein accompany each nerve that penetrates the epimysium of a skeletal muscle. Branches of the nerve and blood vessels follow the connective tissue components of the muscle of a nerve cell and with one or more minute blood vessels called capillaries.

Function of Skeletal Muscle

When you want to move your arm, your brain sends a nervous signal through your nerves. The simple act of raising your arm requires many muscles, so the signal is sent down many nerves to many muscles. Each skeletal muscle receives the nervous impulse at neuromuscular junctions. These are places where nerves can stimulate an impulse in a muscle cell. The impulse travels down channels in the sarcolemma, the plasma membrane of skeletal muscle cells. At certain places in the membrane, there are channels that lead inside the cell. These transverse tubules carry the nervous impulse inside the cell. The impulse releases calcium ions from a specialized endoplasmic reticulum, the sarcoplasmic reticulum. These calcium ions active troponin to release from tropomyosin. Tropomyosin can then shift position, allowing the myosin heads to attach to the actin filament.

Once the myosin heads are attached, the ATP available will be used to contract the filament. This is done by each pair of myosin heads slowly crawling down the filament. Energy from ATP is used to move one head, while the other is attached. When many hundreds or thousands of heads are involved, this quickly contracts the sarcomere up to 70% of its original length. As the nervous impulse hits each muscle fiber and muscle at the same time, the arm can lift in a fluid motion. As an added feedback measure, every skeletal muscle has special sensory cells which send feedback to the brain. These cells, called muscle spindles, have specialized proteins which can sense tension. When tension is received by the cell, the cell starts a nervous impulse and sends the signal through neurons to the brain.

By piecing together this complicated framework of inputs and outputs, the brain can sense where the body is in space. The somatic nervous system controls these actions, and allows us to move our body in a coordinated manner. Skeletal muscle is controlled almost exclusively by the somatic nervous system, while cardiac and smooth muscle is controlled by the autonomous nervous system. This system can be easily demonstrated. Close your eyes, then clap your hands together several times. Did your hands meet? This is because your brain has been training in coordination since birth, and recognizes the specific tensions on each muscle as you swing your hands. As you clap, these inputs are monitored and adjustments are made to ensure your hands continue to make contact with each other. The same system is responsible for balance, coordination, and most physical movements.

Skeletal Muscle Location

Skeletal muscle, as the name implies, is any muscles that connects to and controls the motions of the skeleton. In all there are somewhere between 600 and 900 muscles in the human body, but an exact number is hard. Many muscles are obscurely small or are sometimes grouped together with similar muscles. Skeletal muscle is found between bones, and uses tendons to connect the epimysium to the periosteum, or outer covering, of bone.

Skeletal muscle is adapted and shaped in many different ways, which give rise to complex movements. Skeletons are not always internal as they are in humans. Even animals with exoskeletons, like crabs and mussels, have skeletal muscle. While the muscle might be adapted differently depending on the animal, skeletal muscle is defined by its striations and connections to skeleton. Everything from the flapping of a bird's wings to the crawling of a beetle are carried out by skeletal muscle.

Anatomical terms of Muscle

There is more than one way to categorize the functional role of muscles. It depends on perspective. We may look at the muscles in terms of their function in specific movements or we may look at them in terms of the entire body as a system, complete with many subsystems. The latter view is not what we are concerned with in this explanation but the when viewed this way muscles are classified according to their function rather than their role in a particular movement. The word stabilizer or stabilization, therefore, has a much broader and complex definition.

This view sees the body as a system of motor (or mobilizer) and stabilizer muscles. This concept was first proposed by Rood and furthered by the work of Janda and Sahrmann as well as by Comerford and Mottram who proposed the concept of local and global stabilizers and global mobilizers.

Although, the concept of a stabilizing muscle can still be viewed in terms of a single movement in this system, certain muscles are considered to have the primary function of stabilizers in the body, being, by virtue of their position, shape, angle or structure, more suited to work as a stabilizer than as a mobilizer.

Although its complexities go way beyond the scope of this explanation (and the expertize of its author), this way of looking at the body is a valid and important one for the strength trainee. For instance, this view teaches us that the abdominal group of muscles, once primarily thought of as a muscle we perform situps with, is much more important as a major stabilizer of the spine. This may lead us to train those muscles in a way that supports their function, thus making us stronger and more injury free. This, in fact, is one of the hallmarks of "functional" training, although the term has been much abused and overused.

The type of stabilizer, however, are fixators, which are active during one movement and at one joint. There are certain muscles that act primarily as stabilizes because of their angle of pull. An example of such muscles is a group of muscles known as the rotator cuff muscles of the shoulder girdle. This group comprises the supraspinatus, infraspinatus, teres minor and subscapularis. These muscles are mainly known as muscle of rotation for their contribution to external and internal rotation of the shoulder but they are actually much better suited for the primary role of stabilization and they are very important in stabilizing the humeral head in the glenoid fossa.

Agonist Muscle

Agonist muscles are muscles that are responsible for causing a certain joint motion. However, the term is often defined incorrectly to mean all the muscles that have a role in producing a movement. By this definition stabilizers, neutralizers, and fixators are also agonists. This is incorrect.

An agonist is a muscle that is capable of increasing torque in the direction of a limb's movement and thus produce a concentric action. In other words, the muscle can produce a force that accelerates a limb around its joint, in a certain direction. This does NOT mean that this direction is the only one the muscle can produce force in but only that it is capable of this and thus is directly involved in producing a certain movement, making it a prime mover. To keep it simple, then, an agonist is a muscle that causes rotational movement at a joint by producing torque. A movement can always have more than one agonist although a certain agonist may be capable of producing more torque than its partner. They are also sometimes called protagonists.

Agonist   **Antagonist**

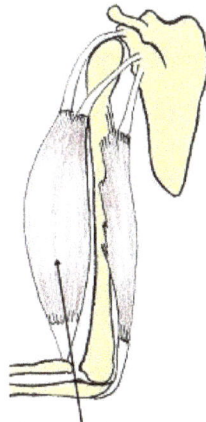

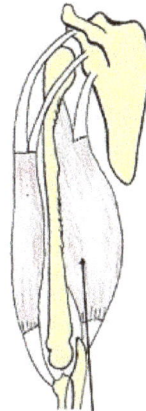

The biceps brachii is the agonist which flexes the elbow.   *The triceps is the antagonist which resists flexion and extends the elbow.*

Agonist and antagonist relationship of biceps and triceps muscle

Many people refer to muscles having a redundant role in producing torque about a joint as being synergistic agonists but with one of these muscles being the prime mover. This is a silly and arbitrary distinction since there are many instances where a muscle with a redundant role can take over for a paralyzed one, making that muscle the "prime mover". Agonist and "prime mover" simply speaking, means the same thing and the terms are interchangeable. However, sometimes it is useful to refer to one muscle, usually a larger one that articulates at more than one joint, as the prime mover. In this way, the prime mover can be spoken of in relation to its fixators or supporters. This type of instance is very common in that certain terms only become useful in a specific context. The biceps brachii, which will be used as an example from here on, is often considered the prime mover in elbow flexion, although it is only one of several flexors of the elbow joint.

The brachialis, for instance, is another elbow flexor, located inferior to the biceps on the upper arm. Unlike the biceps, which inserts onto the radius, which is able to rotate, the brachialis inserts onto the ulna which cannot rotate. This, it can be said that the brachialis is the only pure flexor of the elbow joint whereas the larger biceps can also supinate the forearm.

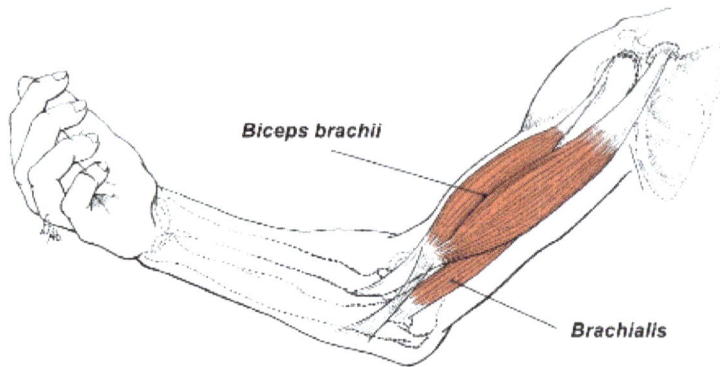

When most people think of elbow flexion they think of the more superficial biceps brachii.
But the brachialis is the only pure elbow flexor

A muscle can only be referred to as an agonist in relation to a movement or another muscle. It is never proper to call any one muscle an agonist unless we are describing its role in a movement or we are referring to it in terms of a muscle on another side of the joint, known as an antagonist. To say "the biceps is an agonist" is incorrect or at least incomplete (which comes down to the same thing).

The biceps brachii is an agonist for elbow flexion. It is assisted by the brachialis and the brachioradialis. These are the agonists of elbow flexion, all of which are capable of flexing the elbow joint to some extent.

Antagonist Muscle

An antagonist is a muscle that is capable of opposing the movement of a joint by producing torque that is opposite to a certain joint action. This is usually a muscle that is located on the opposite side of the joint from the agonist. The triceps, an extensor of the elbow joint, is the antagonist for elbow flexion, and it would also be correct to say that the tricep is an antagonist to the biceps, and vice versa.

In order for an agonist to shorten as it contracts the antagonist must relax and passively lengthen. This occurs through reciprocal inhibition, which is necessary for the desig-nated joint movement to occur unimpeded. Reciprocal inhibition is a neural inhibition of the motor units of the antagonist muscle. When the agonist muscle contracts, this causes the antagonist muscle to stretch. Normally, this stretching would be followed by a stretch reflex which would make the muscle being stretched contract against the change in length. If this were allowed to happen unchecked then it would result in very jerky or oscillatory movement since the stretch reflex in the antagonists would elicit a new stretch reflex in the agonist, so on and so forth. The inhibition of the alpha-moto-neurons in the antagonist are brought about by Ia-inhibitory interneurons of the spinal cord, which are excited by IA afferents in the agonist muscle.

However, antagonists are not always inactive or passive during agonist movements. Antagonists also produce eccentric actions in order to stabilize a limp or decelerate a movement at the end of a motion. For instance, during running the hip extensors are antagonists to the hip flexors, which act to bring the femur forward during the running stride. So, the hip extensor muscles must relax to some degree to allow this forward motion of the thigh to take place. However, the extensors must also act to arrest this forward motion at the top of the stride. So the antagonists both relax to allow the mo-tion to happen and then contract to put the brakes on it. This makes for a very fine balance of activity between agonist and antagonist pairings.

Agonist Antagonist Coactivation or Co-contraction

When both the agonist and antagonist simultaneously contract this is called coactiva-tion. It can be advantageous for coactivation to occur for several reasons. For instance, when movements require a sudden change in direction, when heavy loads are carried, and to make a joint stiffer and more difficult to destabilize.

The purported reason that co-contraction may occur during changes in direction is that modulating the level of activity in one set of muscles is more economical than al-ternately turning them on and off. For heavy loads, increased joint stiffness is desirably for lifting heavier loads and co-contraction of the core muscles of the torso routinely occurs during these activities. For fine motor activities of the fingers, as well, complex co-contraction activity is needed.

Synergist Muscle

A synergist muscle is a muscle which works in concert with another muscle to generate movement. These muscles can work with the so-called agonists or prime movers which surround a joint, or the antagonistic muscles, which move in the opposite direction. For many common movements, from turning the head to pointing the toes, a synergist muscle or group of muscles is required.

Prime movers are designed to move a joint in a particular direction, but a single prime mover or group of prime movers requires a synergist muscle to control the movement. Synergist muscles stabilize muscle movements to keep them even, and they control the movement so that it falls within a range of motion which is safe and desired. One could think of these muscles as helping hands which focus the effort of the movement to create a high level of control. By working synergistically, muscles also reduce the amount of work they need to do, which can improve endurance.

Sometimes, a synergist muscle can form part of what is known as a fixator group. Fixators are designed to "fix" or stabilize a joint. For example, when people stand up, fixator groups at the ankles keep the joints stable so that the ankles will not bend or wobble, causing difficulties with balancing. Fixator groups are also what allow people to isolate movements to a specific joint or area of the body, with the muscle group holding nearby joints in place.

A good example of a pair of synergist muscles can be found in the elbow, where the brachioradialis and biceps work together to extend or flex the arm by moving the elbow joint. Relationships between muscles can also change, depending on the actions involved, with muscles sometimes acting as a synergist muscle, and sometimes working alone or as an antagonist. The biceps and triceps muscles, for example, are considered antagonists because they move the elbow joint in opposite directions.

When people build and tone muscle for physical fitness, they often need to pay attention to working complete muscle groups by moving joints in different ways, to promote even development of muscle, including the synergist muscles which can help to stabilize joints and exert control over muscle movements. By ensuring that muscle groups are worked in a variety of ways, people can improve the strength of all of the muscles which surround a joint, improving range of motion and general fitness in all directions, rather than just one.

Neutralizing Muscles

Neutralizing muscles provide important support during exercise to prevent injury and restrict movement. These muscles are often smaller than the large moving muscles of your body; however, their importance should not be underappreciated or under addressed in a training program. Adequate training of these muscles will improve your overall strength and further reduce the risk of injury.

The contraction of the muscle helps control the movement path of the primary muscle by keeping it within certain definable boundaries. A neutralizers may also be classified as stabilizers because it is their ability to keep joints balanced that holds motion along a specific path. Neutralizers should not be confused with synergistic muscles that help produce the desired movement of the lift. Synergists help control the movement path, but they also help with the primary motion you are trying to achieve.

Example

An examination of the classic biceps curl exercise will help you understand the role of a neutralizer during exercises. During the curl, the angle between your forearm and upper arm closes causing your biceps to contract to produce the movement. However, the bicep also causes supination when contracting. This causes the forearms to rotate so that your palms face upward. To prevent or neutralize this motion from happening, the pronator teres, which is located in your forearm, activates to counter and neutralize this function of your biceps.

Training

Training your neutralizing muscles involves performing exercises in ways that challenge your ability to control the weight your are lifting through different movement types. Performing exercises on a surface that requires stabilization with train both neutralizing and stabilizing muscles related to the specific movement. Since these muscles are smaller and related to postural control, you should use a high number of repetitions during your exercise to elicit maximal training of your neutralizers. Generally, performing at least 20 repetitions of a particular exercise should begin to challenge the neutralizing muscles because your primary muscles will become fatigued and less able to control their movement path.

Benefits

Training the neutralizers will make you strong and faster while preventing injuries. For example, imagine playing a game of soccer and you plant your foot in the ground to make a swift turn. When you plant, your neutralizers act to make sure your primary muscles activate in a way that maximizes the reduction in momentum and production of force to propel your body in the proper direction. Allowing the muscles to apply force unevenly and multidirectionally will result in a loss of force in your desired direction, which will cause you to be slower. The neutralizers of your knee also prevent your knee from moving in an unsafe direction and limit your risk of knee injury during the high-impact cut.

References

- What-is-the-muscular-system-function-how-muscles-work-in-groups: study.com, Retrieved 31 March 2018

- Skeletal-muscle, science: britannica.com, Retrieved 21 April 2018

- Kinesiology: what-is-anagonist-antagonist-stabilizer-fixator: gustrength.com, Retrieved 11 May 2018

- What-is-a-synergist-muscle: wisegeek.com, Retrieved 31 March 2018

- Definition-of-neutralizer-muscles-549731: livestrong.com, Retrieved 19 July 2018

5

Nervous System

The nervous system is an important system in a human body that controls the transmission of signals between different parts of the body. In vertebrates, it consists of two major systems, the peripheral and central nervous systems. The different aspects of the nervous system have been carefully analyzed in this chapter.

The nervous system is a very complex organ system. The nervous system is made up of all the nerve cells in your body. It is through the nervous system that we communicate with the outside world and, at the same time, many mechanisms inside our body are controlled. The nervous system takes in information through our senses, processes the information and triggers reactions, such as making your muscles move or causing you to feel pain. For example, if you touch a hot plate, you reflexively pull back your hand and your nerves simultaneously send pain signals to your brain. Metabolic processes are also controlled by the nervous system.

There are many billions of nerve cells, also called neurons, in the nervous system. The brain alone has about 100 billion neurons in it. Each neuron has a cell body and various extensions. The shorter extensions (called dendrites) act like antennae: they receive signals from, for example, other neurons and pass them on to the cell body. The signals are then passed on via a long extension (the axon), which can be up to a meter long.

The nervous system has two parts, called the central nervous system and the peripheral nervous system due to their location in the body. The central nervous system (CNS) includes the nerves in the brain and spinal cord. It is safely contained within the skull and vertebral canal of the spine. All of the other nerves in the body are part of the peripheral nervous system (PNS).

Regardless of where they are in the body, a distinction can also be made between voluntary and involuntary nervous system. The voluntary nervous system (somatic nervous system) controls all the things that we are aware of and can consciously influence, such as moving our arms, legs and other parts of the body.

The involuntary nervous system (vegetative or autonomic nervous system) regulates the processes in the body that we cannot consciously influence. It is constantly active, regulating things such as breathing, heart beat and metabolic processes. It does this by receiving signals from the brain and passing them on to the body. It can also send signals in the other direction – from the body to the brain – providing your brain with information about how full your bladder is or how quickly your heart is beating, for example. The involuntary nervous system can react quickly to changes, altering processes in the body to adapt. For instance, if your body gets too hot, your involuntary nervous system increases the blood circulation to your skin and makes you sweat more to cool your body down again.

Both the central and peripheral nervous systems have voluntary and involuntary parts. However, whereas these two parts are closely linked in the central nervous system, they are usually separate in other areas of the body.

The involuntary nervous system is made up of three parts:

- Sympathetic nervous system

- Parasympathetic nervous system

- Enteric (gastrointestinal) nervous system

The sympathetic and parasympathetic nervous systems usually do opposite things in the body. The sympathetic nervous system prepares your body for physical and mental activity. It makes your heart beat faster and stronger, opens your airways so you can breathe more easily, and inhibits digestion.

The parasympathetic nervous system is responsible for bodily functions when we are at rest: it stimulates digestion, activates various metabolic processes and helps us to relax. But the sympathetic and parasympathetic nervous systems do not always work in opposite directions; they sometimes complement each other too.

The enteric nervous system is a separate nervous system for the bowel, which, to a great extent, autonomously regulates bowel motility in digestion.

Functions of the Nervous System

The primary function of the nervous system is to receive information, and to generate a response to that stimulus. The information and the response could be simple, subtle or complex. For instance, when a hot object is touched, its temperature is conveyed quickly to the central nervous system and the response is an immediate reflex of removing the hand, through the action of skeletal muscles. A few such incidents could also lead to the formation of learning and long-term memory, encoded as a series of neural connections. Alternatively, it could be the sensation of a cold drink on a hot day, where the body responds with a feeling of pleasure. This is expressed through neuronal activity in various parts of the body, depending on the individual, not relying on any obvious effector cell. On the other end of the spectrum, the stimulus could be indirect, such as the sound of rustling leaves in a quiet forest, indicative of an animal slithering. This could lead to a cascade of responses. The body might respond to this sound with an adrenalin rush, prompting a flight response, and change the metabolic state of skeletal, smooth and cardiac muscles. It could also retrieve memory and try to recollect the possibility of the animal being a venomous snake, and the best possible route for escape. Much of this happens nearly instantaneously. Here, a simple stimuli with a complex response involving effector cells across the body. Some parts of the nervous system can encode information from stimuli so intricately and deeply, that victims of traumatic events relive painful moments in their entirety, with the whole host of physiological responses, even with an unrelated stimulus.

Among the primary modes of input into the nervous system are the electrical impulses that arise from sense organs. Touch, sound, sight, smell and taste are conveyed to the nervous system, in order to integrate information and assess the nature of the external world. Similarly, a number of neurons act as sensors for the internal state of the body. For instance, sensory neurons in the eyes, nose and tongue can inform a person about the presence of delicious food, and create a desire to eat. After the food has been ingested, neurons in the digestive system can sense the stretching of stomach muscles. When this information is conveyed to the central nervous system, it triggers a satiety response – the feeling of 'fullness' and a willingness to stop eating. These are complex responses that do not directly involve only a muscle cell. There is a higher order integration occurring at this point, where memory, learning, cognition and emotional state influence the physiological response mediated by the nervous system.

Thus, while the nervous system can be considered as center for receiving, processing and transmitting information, its functions are complex in most organisms. In humans, it is important for thought, reasoning, language, perception and speech. Parts of the central nervous system control voluntary and involuntary muscle movement, and even peristalsis and digestive movement. It is important for the maintenance of balance, internal temperature regulation, and circadian rhythms. The rate of breathing, blood pressure, and heart rate are also modulated by the nervous system. It integrates its actions with the endocrine system in order to provide the body with a coordinated, and fine-tuned response to stimulus.

Organs of the Nervous System

The nervous system in humans is made of the brain and spinal cord, along with the primary sense organs and all the nerves associated with these organs. The brain and the spinal cord form the central nervous system (CNS). All other neuronal tissue is brought under the umbrella of the peripheral nervous system (PNS). Therefore, the PNS includes neurons within sense organs, other sensory nerves, and all motor nerves that deliver messages to different parts of the body.

Functionally, the organs of the nervous system can be further divided into different parts. For instance, the brain is situated within the cranial cavity and weighs less than 1.5 kgs. However, it is the seat for many higher order mental functions, such as planning, consciousness, perception and language. It is broadly divided into the cerebrum, cerebellum and medulla. The cerebrum is the largest part, and is the section that is seen most obviously in external pictorial representations of the organ. It contains two hemi-spheres of nearly equal size and each hemisphere has four lobes. These lobes, called the parietal, temporal, frontal and occipital, have distinct functions, being involved in impulse control, problem solving, visual perception, hearing, language and speech. Though the hemispheres of the brain have some extent of plasticity, specific tasks re-main localized to specific sections of the cerebral cortex.

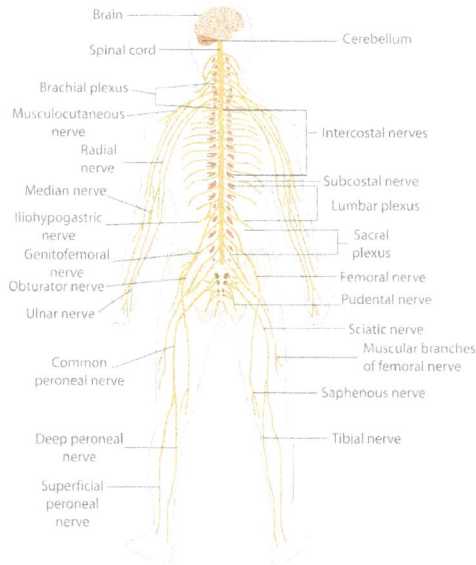

Human nervous system

Neurons form the functional unit of the nervous system. They can be afferent or effer-ent neurons based on whether they carry information towards the CNS or transmit sig-nals from the CNS. Some, called interneurons, are important to integrate information from different stimuli and to create a unified response.

Diseases of the Nervous System

The nervous system can be attacked by infectious pathogens – bacteria, viruses, fungi or protozoans. Bacterial infections such as tuberculosis or syphilis can colonize nervous tissue as a secondary site of infection in advanced stages of the disease. The meninges covering the central nervous system are particularly susceptible to infection, especially when head trauma allows pathogens from other organs access to these delicate tissues, through the cerebrospinal fluid. Other disorders in the nervous system include blocks in the vascular networks of the brain due to strokes. Strokes can lead to large-scale loss of function, up to and including complete paralysis.

Neurons have a very low capacity for regeneration. Therefore, ailments associated with the accumulation of improperly folded proteins are debilitating, since the body can-not completely replace affected cells. Many of these ailments are progressive, i.e., the symptoms become more debilitating with age, and include Alzheimer's disease and

Parkinson's disease. For some ailments, there is a clear genetic factor involved such as in Huntington's disease and in some forms of ataxia. In many such cases, there is a single protein whose gene is mutated in such a way that large-scale changes to the DNA sequence accumulate over successive generations. Therefore, the disease begins to appear earlier and earlier, as misfolded proteins overwhelm cells across the body, until an individual succumbs even before having children.

In most other neurodegenerative diseases, both genetic and environmental factors seem to be involved.

Alzheimer's Disease

The cause for Alzheimer's disease is still not known, though autopsies of patients who have suffered from the ailment often reveal protein plaques in the brain. The earliest hypothesis about the cause of this disease involves a deficiency in a neurotransmitter, and the degeneration of the neurons that depend upon this molecule. Other theories involve specific proteins (amyloid precursor protein, tau protein) that form aggregates in the extracellular space in the brain. There is also some evidence to suggest that the vascular structure of the brain may determine susceptibility to the disease.

Most patients have an enlargement of the brain ventricles, and a shrinking of active nervous tissue in the cortex and hippocampus. Therefore, they show a progressive decline in cognitive function, learning, memory, and mood regulation. Short-term memory loss and the inability to acquire new learning are among the earliest symptoms. Thus, they might repeat themselves frequently, since they cannot remember the contents of the earlier parts of a conversation. As the ailment progresses, they might only retain their earliest memories. They may no longer recognize their caregivers, or remember where they live. There is a simultaneous loss of language as well, and some patients develop paranoia.

Parkinson's Disease

Unlike Alzheimer's disease, Parkinson's does not severely affect cognition. However, there is progressive loss of motor ability, beginning from fine motor skills, and changes to posture and balance. This is usually followed by the appearance of mild tremors, especially in the fingers, or toes. Slowly, there is difficulty in performing repetitive tasks with the hands or legs, such as writing or walking. Deliberate movements become difficult, especially ones that require coordination between the limbs and eyes. The primary region of the brain affected by the disease is the substantia nigra, a region in the midbrain. As in Alzheimer's, the definitive cause for Parkinson's disease is not known. While genetics does play a role, environmental pollutants, injuries or even diet can influence the onset of the disease.

As the disease progresses, the patient needs help for accomplishing every task, including the maintenance of personal hygiene. Since cognitive function is unimpaired, he is aware of his dependency. Therefore, managing the emotional impact of Parkinson's is as important as attending to the loss of motor ability.

Central Nervous System

The central nervous system (CNS) consists of the brain and spinal cord, integrating and coordinating the activities of the entire body. Through these physical organs, thought, emotion, and sensation are experienced, and movement is organized. Long-term and short-term metabolism and homeostasis are regulated through close interaction with the endocrine system.

While the CNS is functionally made of neurons, other cell types such as glial cells play important supportive roles. Some cranial nerves, like the optic and olfactory nerves, are also considered to be a part of the central nervous system.

Functions of the Central Nervous System

The primary function of the central nervous system is integration and coordination. The CNS receives input from a variety of different sources, and implements an appropriate response to the stimuli, in a cohesive manner. For instance, in order to walk the CNS needs visual and integumentary cues – the texture of the surface, its incline, the presence of obstacles, and so forth. Based on these stimuli, the CNS alters skeletal muscle contraction. Once infants learn to walk, this happens involuntarily, no longer requiring conscious thought or concentration. A similar process of receiving complex stimuli and generating a coordinated response is required for vastly varied activities – whether it is balancing a bicycle, maintaining a conversation or mounting an immune response.

The CNS, especially the brain, is considered the physical seat for most higher order mental functions, with neuronal connections forming the basis for thought and retention of memory. The brain plays an important role in the development of speech, language and communication, involving an association of abstract symbols and sounds with concrete objects and emotions. Motivation, ambition, reward and satisfaction are also mediated through neuronal connections in the CNS. At the same time, the limbic system of the brain also controls the most basic emotions and drives, such as pleasure, fear, anger, hunger, thirst, sleepiness and sexual desire. In addition, involuntary reflexes are mediated by the spinal cord, providing protection and quickly preventing injury.

The CNS directly or indirectly influences nearly every internal organ system, whether related to respiration, digestion, excretion, circulation or reproduction.

Example of Central Nervous System Activity

The key to the work of the CNS is integration. It receives input from various sources and creates a cohesive response. This is particularly important for animals in complex social structures, like human beings. For instance, meeting an old friend and catching up over coffee can seem like a relaxing event. However, to facilitate a successful interaction, the CNS needs to be abuzz with activity. It begins when you see the friend and recognize her – your brain is correlating the neurochemical signals received from the optic nerve with the image you have in memory. It proceeds with the recollection of common experiences and the slipping into the vernacular of an earlier time. Some research suggests that the CNS can even associate different body language with different sets of people or events. You may find yourself using phrases that haven't been in your vocabulary for years, or changing your accent and posture slightly, without being actively aware of it. The CNS retrieves memory, correlates with current (the sight of your friend and your conversation) to generate an emotional as well as physiological response. It may end with the brain directing skeletal muscles to walk towards a coffee house, instructing the vocal chords to issue an invitation, and even using your understanding of cultural markers to determine whether a hug or a handshake would be an appropriate end to the meeting.

Anatomy of the Central Nervous System

In vertebrates, the brain and spinal cord are encased in bony cavities, with the brain residing within the skull, and the vertebral column protecting the spinal cord. Three membranous coverings, called the meninges, provide mechanical support and protection to the central nervous system. These meninges are called pia mater, arachnoid mater and dura mater. Pia mater is the layer closest to the nervous tissue and dura mater lies next to the bone. Additionally, cerebrospinal fluid (CSF), produced in the four ventricular cavities of the brain, flows between the pia mater and arachnoid mater, providing protection from pathogens and mechanical support to the entire central nervous system. Special glial cells called ependymal cells produce CSF.

The brain is made of the cerebrum, cerebellum and brain stem. The cerebrum consists of two large hemispheres demarcated by a thick band of nerve fibers called the corpus callosum. Each of the hemispheres can be divided into four lobes – the frontal, parietal, temporal and occipital lobes. Each of these lobes is relatively distinct in function, relating to higher levels of cognition (frontal lobe), somatosensory input (parietal lobe), auditory stimuli (temporal lobe) or visual stimuli (occipital lobe). The localization of function to different lobes was initially discovered in patients with brain damage. Further study has indicated some level of plasticity as well as communication and integration between neurons in different lobes.

The outer layer of the cerebrum is called the cerebral cortex and this is usually pinkish grey in color, and contains neural cell bodies. It can be divided on the basis of function into sensory, motor and association areas as shown in the image below. For instance, the primary sensory cortex receives sensory input from the body as well as from specialized sense organs. The motor areas are involved in control and execution of voluntary motor activities. Association areas are necessary for perception, abstract thinking, and associating new sensory input with memory.

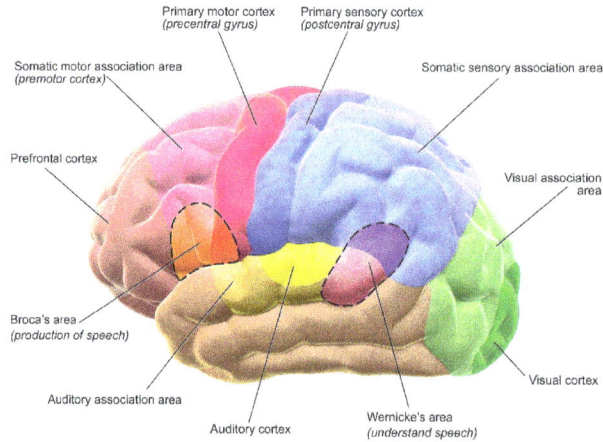

Motor & Sensory of the cerebral cortex

These demarcations of the cerebral cortex are usually represented bilaterally in both hemispheres as seen in the image.

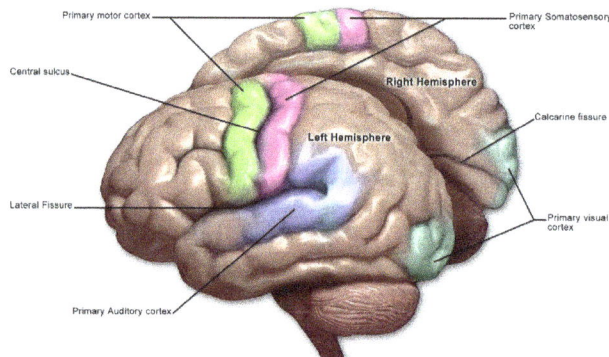

Sensory motor

The cerebellum is smaller than the cerebrum, is made of two lobes, and is located behind the brain stem. It is involved in the coordination of different muscle groups to produce smooth movement, controlling posture and balance. The neurons of the inner ear associated with balance relay their information to the cerebellum, which also receives auditory and visual input.

The brain stem is made of three parts – the midbrain, pons and the medulla oblongata. The medulla controls most involuntary actions, while the midbrain and pons are associated with sensory functions, excitation and motivation. The brain stem connects the brain with the spinal cord.

The spinal cord is about 17 inches in length, tapering along the length of the vertebral column in humans, beginning near the occipital bone and ending at the lumbar region of the spine. It connects the brain with most parts of the body while also containing independent neural networks for pattern generation and for executing reflexes. It can be divided into 31 segments, each giving rise to a pair of spinal nerves. Spinal nerves carry both sensory and motor signals between the body and the spinal cord. The central part of the spinal cord consists of a H-shaped grey column containing the cell bodies of spinal cord neurons. The myelinated axons of these neurons form the white matter.

Brain

The brain is the most complex organ in the human body; the cerebral cortex (the outermost part of the brain and the largest part by volume) contains an estimated 15–33 billion neurons, each of which is connected to thousands of other neurons.

In total, around 100 billion neurons and 1,000 billion glial (support) cells make up the human brain. Our brain uses around 20 percent of our body's total energy.

The brain is the central control module of the body and coordinates activity. From physical motion to the secretion of hormones, the creation of memories, and the sensation of emotion.

To carry out these functions, some sections of the brain have dedicated roles. However, many higher functions — reasoning, problem-solving, creativity — involve different areas working together in networks.

The brain is roughly split into four lobes:

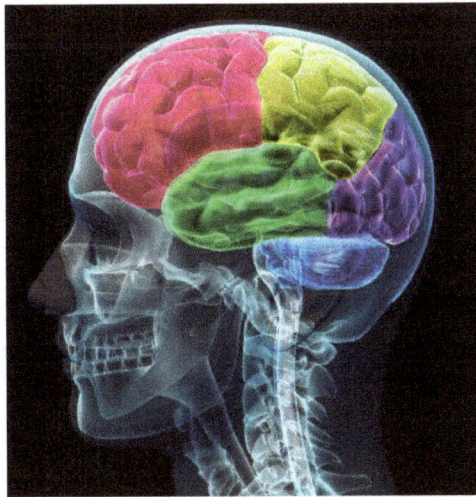

Temporal lobe (green): important for processing sensory input and assigning it emotional meaning.

It is also involved in laying down long-term memories. Some aspects of language perception are also housed here.

Occipital lobe (purple): visual processing region of the brain, housing the visual cortex.

Parietal lobe (yellow): the parietal lobe integrates sensory information including touch, spatial awareness, and navigation.

Touch stimulation from the skin is ultimately sent to the parietal lobe. It also plays a part in language processing.

Frontal lobe (pink): positioned at the front of the brain, the frontal lobe contains the majority of dopamine-sensitive neurons and is involved in attention, reward, short-term memory, motivation, and planning.

Brain Regions

Next, we will look at some specific brain regions in a little more detail:

Basal ganglia: Involved in the control of voluntary motor movements, procedural learning, and decisions about which motor activities to carry out. Diseases that affect this area include Parkinson's disease and Huntington's disease.

Cerebellum: Mostly involved in precise motor control, but also in language and attention. If the cerebellum is damaged, the primary symptom is disrupted motor control, known as ataxia.

Broca's area: This small area on the left side of the brain (sometimes on the right in left-handed individuals) is important in language processing. When damaged, an individual finds it difficult to speak but can still understand speech. Stuttering is sometimes associated with an underactive Broca's area.

Corpus callosum: A broad band of nerve fibers that join the left and right hemispheres. It is the largest white matter structure in the brain and allows the two hemispheres to communicate. Dyslexic children have smaller corpus callosums; left-handed people, ambidextrous people, and musicians typically have larger ones.

Medulla oblongata: Extending below the skull, it is involved in involuntary functions, such as vomiting, breathing, sneezing, and maintaining the correct blood pressure.

Hypothalamus: Sitting just above the brain stem and roughly the size of an almond, the hypothalamus secretes a number of neurohormones and influences body temperature control, thirst, and hunger.

Thalamus: Positioned in the center of the brain, the thalamus receives sensory and motor input and relays it to the rest of the cerebral cortex. It is involved in the regulation of consciousness, sleep, awareness, and alertness.

Amygdala: Two almond-shaped nuclei deep within the temporal lobe. They are involved in decision-making, memory, and emotional responses; particularly negative emotions.

The Brain and Cerebrum

The cerebrum is the largest part of the brain and controls voluntary actions, speech, senses, thought, and memory.

The surface of the cerebral cortex has grooves or infoldings (called sulci), the largest of which are termed fissures. Some fissures separate lobes.

The convolutions of the cortex give it a wormy appearance. Each convolution is delimited by two sulci and is also called a gyrus (gyri in plural). The cerebrum is divided into two halves, known as the right and left hemispheres. A mass of fibers called the corpus callosum links the hemispheres. The right hemisphere controls voluntary limb movements on the left side of the body, and the left hemisphere controls voluntary limb movements on the right side of the body. Almost every person has one dominant hemisphere. Each hemisphere is divided into four lobes, or areas, which are interconnected.

- The frontal lobes are located in the front of the brain and are responsible for voluntary movement and, via their connections with other lobes, participate in the execution of sequential tasks; speech output; organizational skills; and certain aspects of behavior, mood, and memory.

- The parietal lobes are located behind the frontal lobes and in front of the occipital lobes. They process sensory information such as temperature, pain, taste, and touch. In addition, the processing includes information about numbers, attentiveness to the position of one's body parts, the space around one's body, and one's relationship to this space.

- The temporal lobes are located on each side of the brain. They process memory and auditory (hearing) information and speech and language functions.

- The occipital lobes are located at the back of the brain. They receive and process visual information.

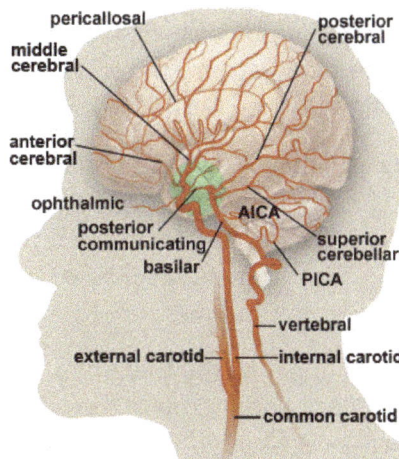

Anatomy of the brain

The cortex, also called gray matter, is the most external layer of the brain and predominantly contains neuronal bodies (the part of the neurons where the DNA-containing cell nucleus is located). The gray matter participates actively in the storage and processing of information. An isolated clump of nerve cell bodies in the gray matter is termed a nucleus (to be differentiated from a cell nucleus). The cells in the gray matter extend their projections, called axons, to other areas of the brain.

Fibers that leave the cortex to conduct impulses toward other areas are termed efferent fibers, and fibers that approach the cortex from other areas of the nervous system are termed afferent (nerves or pathways). Fibers that go from the motor cortex to the brainstem (for example, the pons) or the spinal cord receive a name that generally reflects the connections (that is, corticopontine tract for the former and corticospinal tract for the latter). Axons are surrounded in their course outside the gray matter by myelin, which has a glistening whitish appearance and thus gives rise to the term white matter.

Cortical areas receive their names according to their general function or lobe name. If in charge of motor function, the area is called the motor cortex. If in charge of sensory function, the area is called as sensory or somesthetic cortex. The calcarine or visual cortex is located in the occipital lobe (also termed occipital cortex) and receives visual input. The auditory cortex, localized in the temporal lobe, processes sounds or verbal input. Knowledge of the anatomical projection of fibers of the different tracts and the relative representation of body regions in the cortex often enables doctors to correctly locate an injury and its relative size, sometimes with great precision.

Central Structures of the Brain

The central structures of the brain include the thalamus, hypothalamus, and pituitary gland. The hippocampus is located in the temporal lobe but participates in the processing of memory and emotions and is interconnected with central structures. Other structures are the basal ganglia, which are made up of gray matter and include the amygdala (localized in the temporal lobe), the caudate nucleus, and the lenticular nucleus (putamen and globus pallidus). Because the caudate and putamen are structurally similar, neuropathologists have coined for them the collective term striatum.

- The thalamus integrates and relays sensory information to the cortex of the parietal, temporal, and occipital lobes. The thalamus is located in the lower central part of the brain (that is, upper part of the brainstem) and is located medially to the basal ganglia. The brain hemispheres lie on the thalamus. Other roles of the thalamus include motor and memory control.

- The hypothalamus, located below the thalamus, regulates automatic functions such as appetite, thirst, and body temperature. It also secretes hormones that stimulate or suppress the release of hormones (for example, growth hormones) in the pituitary gland.

- The pituitary gland is located at the base of the brain. The pituitary gland produces hormones that control many functions of other endocrine glands. It regulates the production of many hormones that have a role in growth, metabolism, sexual response, fluid and mineral balance, and the stress response.

- The ventricles are cerebrospinal fluid-filled cavities in the interior of the cerebral hemispheres.

Base of the Brain

The base of the brain contains the cerebellum and the brainstem. These structures serve complex functions. Below is a simplified version of these roles:

- Traditionally, the cerebellum has been known to control equilibrium and coordination and contributes to the generation of muscle tone. It has more recently become evident, however, that the cerebellum plays more diverse roles such as participating in some types of memory and exerting a complex influence on musical and mathematical skills.

- The brainstem connects the brain with the spinal cord. It includes the midbrain, the pons, and the medulla oblongata. It is a compact structure in which multiple pathways traverse from the brain to the spinal cord and vice versa. For instance, nerves that arise from cranial nerve nuclei are involved with eye movements and exit the brainstem at several levels. Damage to the brainstem can therefore affect a number of bodily functions. For instance, if the corticospinal tract is injured, a loss of motor function (paralysis) occurs, and it may be accompanied by other neurologic deficits, such as eye movement abnormalities, which are reflective of injury to cranial nerves or their pathways in the brainstem.

  o The midbrain is located below the hypothalamus. Some cranial nerves that are also responsible for eye muscle control exit the midbrain.

  o The pons serves as a bridge between the midbrain and the medulla oblongata. The pons also contains the nuclei and fibers of nerves that serve eye muscle control, facial muscle strength, and other functions.

  o The medulla oblongata is the lowest part of the brainstem and is interconnected with the cervical spinal cord. The medulla oblongata also helps control involuntary actions, including vital processes, such as heart rate, blood pressure, and respiration, and it carries the corticospinal (that is, motor function) tract toward the spinal cord.

Spinal Cord

The spinal cord is a complex cylinder of nerves that starts at the base of your brain and runs down the vertebral canal to the backbone. It is part of the body's collection of nerves, called the central nervous system, along with the brain. In each of the spinal cord's many segments lives a pair of roots that are made up of nerve fibers. These roots are referred to as the dorsal (which is towards the back) and the ventral (which is away from the back) roots.

We depend on the spinal column for the main support of our body. It allows us to stand upright, bend, and twist, while protecting the spinal cord from injury. If the spinal cord is injured it often causes permanent changes in the body's strength, sensation, and a handful of other functions due to it's connection to the brain.

Because the spinal cord is the center of the body's functions, a person's life can be drastically changed when an injury is severe enough. There is a lot of research being done for the treatment of spinal cord injuries and scientists are optimistic that the advances they are finding will eventually be enough to fully repair damages. In order to understand how a spinal cord injury can affect a person's life, you will need a good handle on the multiple functions that a spinal cord provides.

Major Functions of the Spinal Cord

The spinal cord's major functions include:

- Electrical communication: Electrical currents travel up and down the spinal cord, sending signals which allow different segments of the body to communicate with the brain.

- Walking: While a person walks, a collection of muscle groups in the legs are constantly contracting. The action of taking step after step may seem incredibly simple to us since we have been doing it all of our lives, but there are actually a lot of factors that have to be coordinated properly to allow this motion to occur. This central pattern generators in the spinal cord are made up of neurons which send signals to the muscles in the legs, making them extend or contract, and produce the alternating movements which occur when a person walks.

- Reflexes: Reflexes are involuntary responses resulting from stimuli involving the brain, spinal cord, and nerves of the peripheral nervous system.

Structure of the Spinal Cord

The overall structure of the spinal cord is enclosed by the protection of the vertebral column. The spinal nerves are located in the spaces between the arches of the vertebrae. Spinal nerves are divided into these separate regions:

- Cervical (neck)
- Thoracic (chest)

- Lumbar (abdominal)

- Sacral (pelvic)

- Coccygeal (tailbone)

White Matter and Grey Matter

The spinal cord is split into grey matter (which is in the shape of a butterfly) and white matter (which is the material surrounding the grey). The white matter is made up of nerve fibers, called axons, which run up and down the length of the cord. Each group of axons carries a specific type of information it needs to communicate. Ascending tracts of axons communicate with the brain, while the descending carry signals from the brain to various muscles and glands throughout the body.

The grey matter is also arranged according to it's function. If you were to split the grey matter into two halves: each half has a dorsal horn, ventral horn, and a lateral horn. The dorsal and ventral horns supply skeletal muscle, while the lateral horn supplies cardiac and smooth muscle.

Spinal Nerves

Spinal nerves are what allow the spinal cord and the rest of the body to communicate. A nerve is an organ shaped like a small cord that is made up of several axons that are bound together. There are 31 pairs of spinal nerves:

- 8 are cervical nerves located in the neck

- 12 are thoracic nerves located in the chest

- 5 are lumbar nerves located in the abdomen

- 5 are sacral nerves located in the pelvis

- 1 is the coccygeal nerve located in the tailbone

Reflexes

A reflex can be a simple and uncontrolled response or a learned response. The simple ones are built into our nervous system, such as pulling your hand away from something hot. A reflex that is acquired comes from practice, such as playing the piano. A reflex is made up of 5 components:

1. Receptor: The receptor responds to an electrical signal.

2. Afferent pathway: This pathway sends the action onto the integrating centre.

3. Integrating centre: This is typically the nervous system and is where all of the action potentials are processed. Once the information is processed the integrating centre determines how the body should respond.

4. Efferent pathway: The response then travels through this pathway to the effector organ.

5. Effector organ: This organ carries out the response to all of the above. The organ responding is usually a muscle or gland in the body.

Spinal Cord Injury

A spinal cord injury (SCI) is when a part of the cord or the nerves located at the base of the spine are damaged. This can have a major effect on the body's "sensory, motor, and reflex capabilities if the brain is unable to send information past the location of the injury."

The closer the injury is to the brain, the more expansive the damage. As you can probably imagine, an SCI can alter a person's life forever. However, there are many options for treatment available and research results for a paralysis cure have never been more promising.

Peripheral Nervous System

The peripheral nervous system (PNS) consists of all neurons that exist outside the brain and spinal cord. This includes long nerve fibers containing bundles of axons as well as ganglia made of neural cell bodies. The peripheral nervous system connects the central nervous system (CNS) made of the brain and spinal cord to various parts of the body and receives input from the external environment as well.

Functionally, the PNS is divided into sensory (afferent) and motor (efferent) nerves, depending on whether they bring information to the CNS from sensory receptors or carry instructions towards muscles, organs or other effectors. Motor nerves can be further classified as somatic or autonomic nerves, depending on whether the motor activity is under voluntary conscious control.

Anatomically, the PNS can be divided into spinal and cranial nerves, depending on whether they emerge from the spinal cord or the brain and brainstem. Both cranial and spinal nerves can have sensory, motor or mixed functions. The enteric nervous system, surrounding the gastrointestinal tract is another important part of the peripheral nervous system. While it receives signals from the autonomic nervous system, it can function independently as well and contains nearly five times as many neurons as the spinal cord.

Functions of the Peripheral Nervous System

The primary function of the peripheral nervous system is to connect the brain and spinal cord to the rest of the body and the external environment. This is accomplished through nerves that carry information from sensory receptors in the eyes, ears, skin, nose and tongue, as well as stretch receptors and nociceptors in muscles, glands and other internal organs. When the CNS integrates these varied signals, and formulates a response, motor nerves of the PNS innervate effector organs and mediate the contraction or relaxation of skeletal, smooth or cardiac muscle.

Thus, the PNS regulates internal homeostasis through the autonomic nervous system, modulating respiration, heart rate, blood pressure, digestion reproduction and immune responses. It can increase or decrease the strength of muscle contractility across the body, whether it is sphincters in the digestive and excretory systems, cardiac muscles in the heart or skeletal muscles for movement. It is necessary for all voluntary action, balance and maintenance of posture and for the release of secretions from most exocrine glands. The PNS innervates the muscles surrounding sense organs, so it is involved in chewing, swallowing, biting and speaking. At the same time, it mediates the response of the body to noxious stimuli, quickly removing the body from the injurious stimulus, whether it is extremes in temperature, pH, or pressure, as well as stretching and compressing forces.

Examples of the Peripheral Nervous System Response

In a dimly lit room, the pupils of the eye are enlarged, to allow maximum light to fall on the retina. When a bright light is suddenly turned on, sensory receptors in the eye communicate this to the CNS. The response to this new stimulus is mediated through the peripheral nervous system, by contracting the pupils, using external eye muscles to squint and probably even moving the skeletal muscles of the arm to shield the eye.

Similarly, when a sharp or pointed object is stepped on, pain and stretch receptors in the skin send signals to the CNS, which immediately brings about a change in posture, and balance, protecting the foot from potential injury.

Anatomy of the Peripheral Nervous System

The peripheral nervous system is made of nerves, ganglia and plexuses. A nerve contains the axons of multiple neurons bound together by connective tissue. The axon itself is often myelinated, containing a phospholipid secreted by a glial cell called the Schwann cell. The thin covering of Schwann cell cytoplasm forms the innermost layer protecting an axon and is called the neurilemma or neurolemma.

Nerve Structure

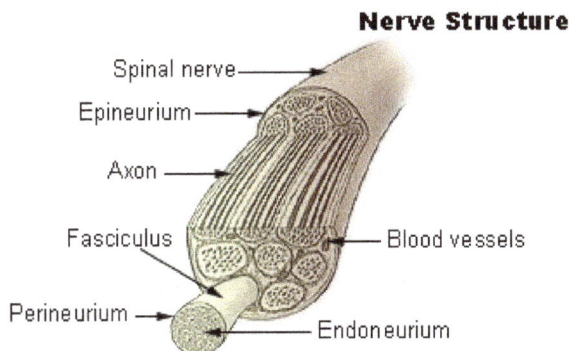

Spinal nerve
Epineurium
Axon
Fasciculus
Perineurium
Blood vessels
Endoneurium

The image above depicts the structure of a nerve. Blood capillaries and other connective tissue around the neurilemma form the endoneurium. When multiple axons are bundled together to form structures called fascicles, a fibrous tissue called the perineurium holds them together. Finally, the whole nerve containing numerous axon bundles is encased in fibrous epineurium. The cell bodies or soma of these neurons also cluster together and are covered by the epineurium to form ganglia that look like swellings on the nerve fiber. In the autonomic nervous system, these ganglia become the sites for synaptic transmission between two neurons. Branching networks of intersecting spinal and autonomic nerves form structures called plexuses that have both sensory and motor functions and serve a particular region of the body.

The PNS can be said to consist of 12 pairs of cranial nerves and 31 pairs of spinal nerves. Cranial nerves emerge in pairs on either side of the base of the skull, through small openings called foramina. Cranial nerves are numbered using roman numerals I-XII, depending on their position while exiting the cranium. A potentially vestigial nerve called cranial nerve zero emerges anterior to the first cranial nerve. Cranial nerves also have a Latin or Greek name, based on their structure or their effector organ. They primarily innervate the head and neck, with the significant exception of the tenth cranial nerve, also known as the vagus nerve. Some cranial nerves have only sensory function, such as the olfactory and optic nerves. The structure of these nerves also occasionally leads to their classification under the central nervous system. Cranial nerve VIII is another sensory nerve relating to hearing and balance. Motor nerves contain nerve fibers that carry signals to muscles of the pupil of the eye or external eye muscles. The rest are mixed nerves containing both sensory and motor nerve fibers. Among these, the XI and XII cranial nerves mostly serve a motor function, and innervate the neck, back and tongue. The vagus nerve is another mixed nerve that carries signals from internal organs to the brain and conducts impulses to the organs of the thorax, abdomen and respiratory muscles of the pharynx and larynx. It plays an important role in the parasympathetic innervation of the body.

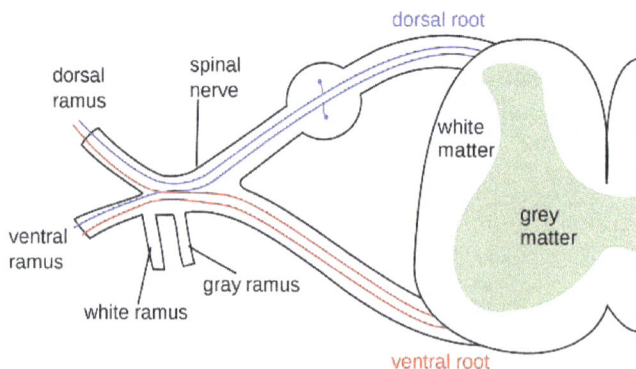

There are 31 pairs of spinal nerves, arising from different regions of the spinal cord. There are 8 that emerge from the cervical region, 12 from the thoracic region, 5 each from the lumbar and sacral regions and 1 pair of spinal nerves from the coccygeal region. Each spinal nerve is a mixed nerve formed by a combination of afferent and efferent neurons.

The image shows the region near the spinal cord, where every spinal nerve has a posterior and anterior root. The anterior or ventral root contains motor neurons while the dorsal or posterior root has ganglia containing the cell bodies of afferent sensory neurons. Distal to the spine, the nerve again splits into an anterior and posterior ramus in addition to forming a small meningeal branch. The posterior ramus leads to the muscles, joints and skin on the back. The anterior ramus is involved in innervating the skin and muscles of the trunk and leads towards the limbs. Often, the anterior ramus forms a network of intersecting nerve fibers to create plexuses.

Somatic Nervous System

The somatic system is the part of the peripheral nervous system that is responsible for carrying motor and sensory information both to and from the central nervous system. This system is made up of nerves that connect to the skin, sensory organs, and all skeletal muscles. The system is responsible for nearly all voluntary muscle movements as well as for processing sensory information that arrives via external stimuli including hearing, touch, and sight.

Whether you want to learn ballet, throw a ball, or go for a jog, the somatic nervous system plays a vital role in initiating and controlling the movements of your body. How exactly does this complex system work? Let's start by taking a closer look at the key parts of the somatic nervous system.

Parts of the Somatic Nervous System

The term "somatic nervous system" is itself drawn from the Greek word *soma*, which means "body," which is appropriate considering it is this system that transmits the information to and from the CNS to the rest of the body.

The somatic nervous system contains two major types of neurons:

1. Sensory neurons, also known as afferent neurons, are responsible for carrying information from the nerves to the central nervous system.

2. Motor neurons, also known as efferent neurons, are responsible for carrying information from the brain and spinal cord to muscle fibers throughout the body.

The neurons that make up the somatic nervous system project outwards from the central nervous system and connect directly to the muscles of the body, and carry signals from muscles and sensory organs to the central nervous system.

The body of the neuron is located in the CNS, and the axon then projects and terminates in the skin, sense organs, or muscles.

Reflex Arcs and the Somatic Nervous System

In addition to controlling voluntary muscles movements, the somatic nervous system is also associated with involuntary movements known as reflex arcs. During a reflex arc, muscles move involuntarily without input from the brain.

This occurs when a nerve pathway connects directly to the spinal cord. Some examples of reflex arcs include jerking your hand back after accidentally touching a hot pan or an involuntary knee jerk when your doctor taps on your knee.

You don't have to think about doing these things. Sensory nerves carry signals to the spinal cord, often connect with interneurons in the spine, and then immediately transmit signals down the motor neurons to the muscles that triggered the reflex. Reflex arcs that impact the organs are called autonomic reflex arcs while those that affect the muscles are referred to as somatic reflex arcs.

An example of the Somatic System in Action

The primary function of the somatic nervous system is to connect the central nervous system to the body's muscles and control voluntary movements and reflex arcs. Information taken in by sensory systems is transmitted to the central nervous system. The CNS then sends signals via the nerve networks of the somatic system to the muscles and organs.

For example, imagine that you are out for a jog in the park one brisk winter morning. As you run, you spot a patch of slick looking ice on the path ahead. Your visual system perceives the icy patch and relays this information to your brain. Your brain then sends signals to engage your muscles to take action. Thanks to your somatic system, you are able to turn your body and move to a different part of the path, successfully avoid the icy patch and prevent a possibly dangerous fall on the hard pavement.

Autonomic Nervous System

The autonomic nervous system (ANS) is a complex set of neurons that mediate internal homeostasis without conscious intervention or voluntary control. Cells of the ANS innervate all viscera and influence their activity locally as well as mediate global changes to the metabolic state of the organism. The ANS maintains blood pressure, regulates the rate of breathing, influences digestion, urination, and modulates sexual arousal.

Anatomy of the Autonomic Nervous System

The autonomic nervous system contains two types of neurons that interact with each other at ganglia near the spinal cord. The initial preganglionic neurons begin at the central nervous system in different parts of the spinal cord. These preganglionic neurons form synapses with postganglionic neurons at ganglia that decorate either side of the spinal cord. The postganglionic neuron forms a synapse with effector cells.

There are two main branches to the ANS – the sympathetic nervous system and the parasympathetic nervous system. The neurons of the sympathetic nervous system emerge from the thoracic and lumbar regions of the spinal cord, while the parasympathetic neurons are associated with the cranial and sacral regions. The sympathetic nervous system is usually activated in response to emergencies, especially those that threaten survival. On the other hand, the parasympathetic response is related to enhancing growth and reproduction.

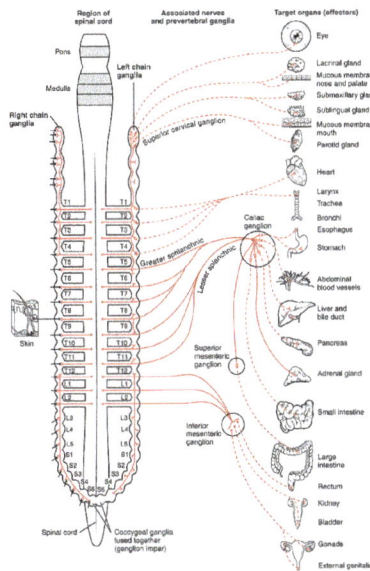

Autonomic Nervous System

Functions of the Autonomic Nervous System

The autonomic nervous system controls the cardiovascular system. It can alter the force and rate of heart contractility, as well as the constriction and dilation of blood vessels. Therefore, it also influences blood pressure. The rate of breathing can also be changed by the ANS. It affects both skeletal and smooth muscle fibers across the body, whether it is changing the metabolism of glucose in skeletal muscles or causing pupil dilation in the eye. The ANS can influence digestive efficiency, altering the secretion of enzymes from glands and the rate of peristaltic movement. For instance, activation of the sympathetic nervous system slows down digestion and diverts blood flow towards skeletal muscle. It can impair sexual arousal and shut down most non-essential functions of the body. On the other hand, the parasympathetic nervous system enhances digestive secretions, peristaltic movements, encourages normal cycles of circadian activity, deep sleep and activates the repair mechanisms of the body. In most cases, a physiological response by the parasympathetic nervous system is in direct opposition to the results mediated by the sympathetic nervous system. Colloquially, the sympathetic nervous system is said to influence the fight-or-flight response, and the parasympathetic nervous system is related to the feed-and-breed, or rest-and-digest responses.

Involuntary actions like sneezing, swallowing or vomiting are also controlled by the ANS. There is evidence that the autonomic nervous system not only influences sexual arousal, but also plays a crucial role in maintaining pregnancy and inducing labor. Finally, the autonomic nervous system also changes urinary output and frequency of micturition.

Examples of the Autonomic Nervous System Response

The autonomic nervous system is often described using the response to imminent physical danger and the recovery of the body after the threat has receded. For instance, when faced with a predator, the body increases heart rate and breathing, reduces digestive secretions and activity, and preferentially diverts blood towards skeletal muscles to enable the body to physically combat the challenge. This is usually accompanied by piloerection to conserve body heat. This is why the sympathetic nervous system is said to mediate the fight-or-flight response. Once the situation has become calmer, the parasympathetic nervous system restores the body towards normal functioning, resuming digestion and excretion, reducing blood pressure and restoring normal circadian rhythms.

However, even in the absence of an external threat, the two branches of the autonomic nervous system undergo changes, and interact closely with the endocrine system to minutely monitor the internal and external environment. For instance, sympathetic activation can lead to an increase in circulating plasma levels of epinephrine and norepinephrine secreted from the adrenal gland. On the other hand, hormones can alter the ANS response as well. In fertile, reproducing mammalian females, this interaction between the ANS and the endocrine systems is particularly interesting. Estrogen is involved in increasing the activity of a crucial part of the parasympathetic nervous system – the vagus nerve. Estrogen simultaneously damps sympathetic nervous system activity while progesterone appears to have the opposite effect. This can be seen in the basal level of heart rate variability (HRV). Usually, heart rate increases during inspiration and decreases during expiration. This variation is normal and is influenced by the vagus nerve. When heart rate variability decreases, it indicates reduced parasympathetic activity. In the follicular phase, under the influence of increased plasma estrogen concentrations, there seems to be an increase in parasympathetic nervous activity affecting HRV. On the other hand, during the luteal phase, HRV points towards a decrease in vagal activity, and a shift in the sympathovagal balance. The importance of these changes to the cardiovascular microenvironment is not fully understood, but it is hypothesized that this could explain the differences in the risk faced by men and women for heart disease. However, it is important to note that gross cardiovascular parameters such as blood pressure or heart rate remain mostly unaffected by the phase of the menstrual cycle due to other compensatory mechanisms.

The interaction between the ANS and cardiovascular system becomes even more important during pregnancy as there are large-scale changes to hemodynamics. Blood volume, basal oxygen consumption, red cell mass, cardiac output, and the heart rate increase in pregnant women. Both systolic and diastolic blood pressure drop and there is extensive remodeling of all blood vessels. While the changing hormonal environment primarily mediates these changes, the ANS is also an important player. Here again, HRV becomes a relatively sensitive and non-invasive measure of ANS activity. Studying the variability in heart rate of pregnant women at different gestational ages points towards an increase in vagal activity in the first trimester, coupled with a decrease in sympathetic nervous system activation. This reverses as gestational age increases, with great spikes in neural activity of the sympathetic nervous system and the release of adrenal hormones – both from the cortex and medulla – as the woman nears term.

After birth, lactation is influenced mostly by the sympathetic nervous system (SNS), since there is very little evidence of parasympathetic innervation of mammary glands. While hormones like oxytocin are important for the stimulation of milk production, activation of the SNS can alter the response of the body to the hormone. Norepinephrine can inhibit blood flow to the mammary glands, inhibit their response to oxytocin, as well as directly influence the release of the hormone from the central nervous system. While these effects have been better studied in cows, there is anecdotal evidence to support this interrelationship in humans as well.

References

- Central-nervous-system: biologydictionary.net, Retrieved 22 May 2018

- Anatomy-of-the-central-nervous-system: emedicinehealth.com, Retrieved 2- March 2018

- Functions-of-the-spinal-cord-what-you-need-to-know: spinalcord.com, Retrieved 11 July 2018

- Peripheral-nervous-system: biologydictionary.net, Retrieved 26 June 2018

- What-is-the-somatic-nervous-system-2795866: verywellmind.com, Retrieved 19 April 2018

- Autonomic-nervous-system: biologydictionary.net, Retrieved 29 May 2018

6

Cardiovascular System

The human cardiovascular system consists of the heart, blood and blood vessels. Arteries, veins and capillaries are responsible for the circulation of blood throughout the human body. This chapter has been designed to provide an elaborate understanding of the cardiovascular system, circulation pathways, the heart and the cardiac cycle.

The cardiovascular system or circulatory system is a system which moves nutrients, gases and wastes between cells, helps fight diseases, and transports blood throughout the body. The main components of the human cardiovascular system include the heart, blood, and various blood vessels. There are several different circuits contained in the cardiovascular system. One of these systems is the pulmonary circuit, which is a "loop", in which oxygenated blood travels through the lungs. The other circuit is the systemic circuit which transports the rest of the blood in a loop through the body. In the cardiovascular system within the heart, there is the cardiac cycle, which is the flow of blood between heartbeats. The cardiovascular system is essential to the human body for blood to be distributed properly.

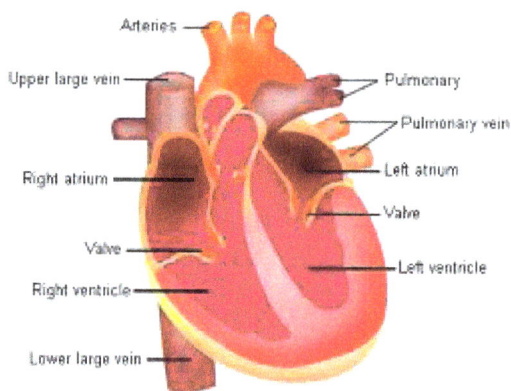

The first major component of the cardiovascular system is the pulmonary circuit. The pulmonary circuit carries deoxygenated blood away from the heart and returns the blood in an oxygenated form. In the heart, there are four chambers, the right and left atria, as well as the right and left ventricles. In pulmonary circulation deoxygenated blood starts in the right side of the heart. The blood is then pumped by the right ventricle of the heart as deoxygenated blood into the pulmonary artery, which is a blood vessel that's function, is to carry blood away from the right ventricle. From the pulmonary artery, the blood is taken into the capillaries, which are small blood vessels that connect arteries and veins of the lungs. Upon the deoxygenated blood entering the lungs, oxygen binds to the red blood cells as carbon dioxide diffuses. Once the blood has become oxygenated it travels along the pulmonary vein into the left atrium of the heart. Upon entering the left atrium the blood joins the systemic circuit.

The systemic circuit is far larger than the pulmonary circuit and therefore is of far greater importance to the body. The main function of the systemic circuit is to carry blood to and away from all the tissues in the body. Due to the fact that this system is much larger, this causes the walls of the left ventricle to be far stronger than the right side of the heart. These thick muscles of the left side of the heart are essential for blood to be distributed throughout the various tissues in the body. In the systemic circuit oxygenated blood is transported from the left ventricle into the aorta, which is the largest artery in the body. The aorta branches downward to carry blood to the respective parts of the body. However, as blood is carried down the aorta, there are several smaller arteries which branch off the aorta to carry blood to their respective parts of the body. As blood exits the left ventricle, and begins to travel down the aorta, small arteries branch off of the aorta to carry blood to the upper torso as well as the brain. When blood travels to the brain, the arteries branch continuously, eventually becoming capillaries which reach every single cell within the brain. When capillaries reach each cell in the brain this allows for every cell in the body to have a supply of oxygen as well as a way of disposing of carbon dioxide. Once the carbon dioxide is disposed from the cells the blood becomes deoxygenated once again. This deoxygenated blood travels to the heart by going through the various veins of the body. These veins which are returning to the heart eventually all link together to form the inferior and superior vena cava, which are the two largest veins in the body. The inferior and superior vena cava pump the deoxygenated blood back into the right atrium, which starts the circulatory process all over again.

Another essential part of the cardiovascular system is the cardiac cycle. The cardiac cycle is measured on the basis of systole and diastole. Systole is the period in which the heart is busy pumping blood. Diastole is the period in which the heart is resting as well as filling up with blood. In the atria of the heart, systole is stimulated by "electric" nerve impulses, which is created by a section on the wall of the right atrium called the sinoatrial node, or (SA). This process is known as atrial systole. The SA is essential as it sets the tone of the beating of the heart, and is therefore referred to as a pacemaker. After atrial systole occurs, the electric impulse travels through the walls of the atria to the atrioventricular node, or (AV), which is located in the wall between the right atrium and right ventricle. After reaching the AV, the impulse signal is delayed 0.1 seconds in order to ensure that the atria of the heart have finished contracting. After this delay, the impulse travels through special fibres in the heart known as Purkinje fibres. Once the electrical impulse reaches the end of the Purkinje fibres, the impulse spreads through the original cardiac muscle, which then allows for ventricular systole to occur. Overall, the cardiac cycle takes about 0.8 seconds to complete. Around half of this time is spent between atrial and ventricular systole, and the other half is composed of atrial and ventricular diastole.

The second part of the cardiac cycle is also a major component of the cardiovascular system. This part of the cycle explains that the "lub-dup" heart sounds are caused by the closing of the valves in the heart during the cardiac cycle. The "lub" sounds is caused by the closing of the AV valves and the beginning of ventricular systole. The "dup" sound is caused by the closing of the pulmonary and aortic valves during ventricular diastole. These valves of the heart ensure that the blood flow in the heart is maintained at the same rate by opening and closing systematically. The cardiac cycle is one of the major components to the cardiovascular system.

The cardiovascular system has shown that it is one of the vital systems of the human body. It is truly amazing how the cardiovascular system is an intricate group of systems which combine together to form one larger system. The cardiovascular system is extremely complicated and the slightest malfunction can cause serious problems in the body. For example, in the heart if the valves are opening the wrong way, a person can have a heart murmur, which in some cases can prove to be fatal. These systems such as the systemic circuit, pulmonary circuit, and cardiac cycle must always be functioning at a very high rate. The cardiovascular system is a magnificent and exciting system to learn.

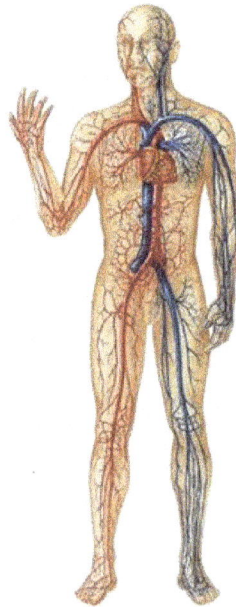

The four major functions of the cardiovascular system are:

1. To transport nutrients, gases and waste products around the body,

2. To protect the body from infection and blood loss,

3. To help the body maintain a constant body temperature ('thermoregulation'),

4. To help maintain fluid balance within the body.

Transportation of Nutrients, Gases and Waste Products

The cardiovascular system acts as an internal road network, linking all parts of the body via a system of highways (arteries and veins), main roads (arterioles and venules) and streets, avenues and lanes (capillaries).

This network allows a non-stop courier system (the blood) to deliver and expel nutrients, gases, waste products and messages throughout the body.

Nutrients such as glucose from digested carbohydrate are delivered from the digestive tract to the muscles and organs that require them for energy.

Hormones (chemical messengers) from endocrine glands are transported by the cardiovascular system to their target organs, and waste products are transported to the lungs or urinary system to be expelled from the body.

The cardiovascular system works in conjunction with the respiratory system to deliver oxygen to the tissues of the body and remove carbon dioxide.In order to do this effectively the cardiovascular system is divided into two circuits, known as the pulmonary circuit and the systemic circuit.

The pulmonary circuit is made up of the heart, lungs, pulmonary veins and pulmonary arteries. This circuit pumps deoxygenated (blue) blood from the heart to the lungs where it becomes oxygenated (red) and returns to the heart.

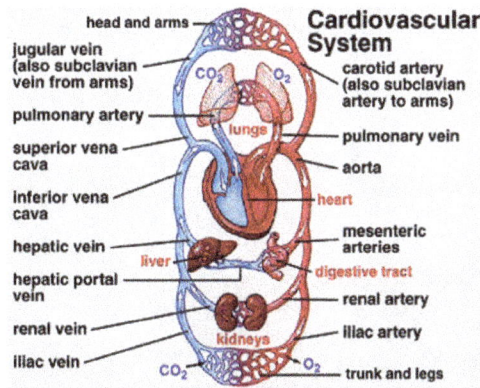

Cardiovascular System

The systemic circuit is made up of the heart and all the remaining arteries, arterioles, capillaries, venules, and veins in the body.

This circuit pumps oxygenated (red) blood from the heart to all the tissues, muscles and organs in the body, to provide them with the nutrients and gases they need in order to function.

After the oxygen has been delivered the systemic circuit picks up the carbon dioxide and returns this in the now deoxygenated (blue) blood, to the lungs, where it enters the pulmonary circuit to become oxygenated again.

Protection from Infection and Blood Loss

Blood contains three types of cells as listed below and shown in the below image:

1.   Red blood cells

2.   White blood cells

3.   Platelets

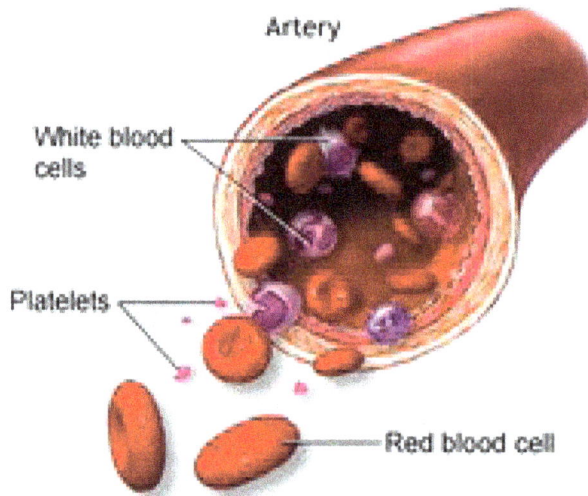

Red blood cells are responsible for transporting oxygen around the body to the tissues and organs that need it.

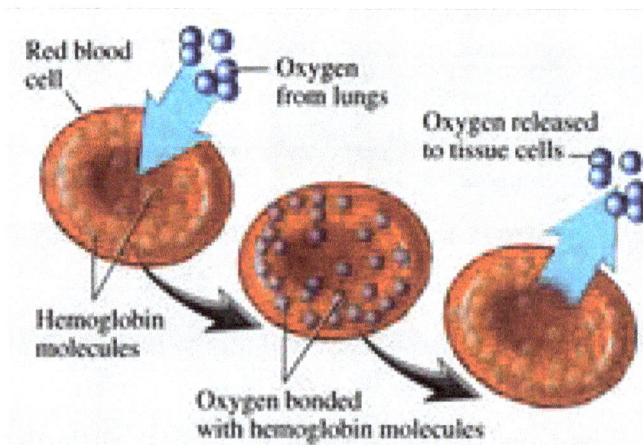

As oxygen enters the blood stream through the alveoli of the lungs it binds to a special protein in the red blood cells called haemoglobin. This can be seen in the above image.

The job of white blood cells is to detect foreign bodies or infections and envelop and kill them.

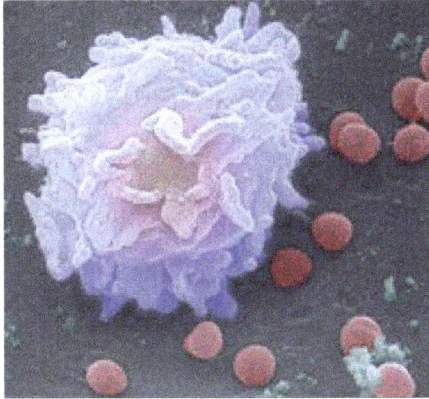

When they detect and kill an infection, they create antibodies for that particular infection which enables the immune system to act more quickly against foreign bodies or infections it has come into contact with previously.

Platelets are cells which are responsible for clotting the blood, they stick to foreign particles or objects such as the edges of a cut.

Platelets connect to fibrinogen (a protein which is released in the site of the cut) producing a clump that blocks the hole in the broken blood vessel. On an external wound this would become a scab.

If the body has a low level of platelets then clotting may not occur and bleeding can continue.

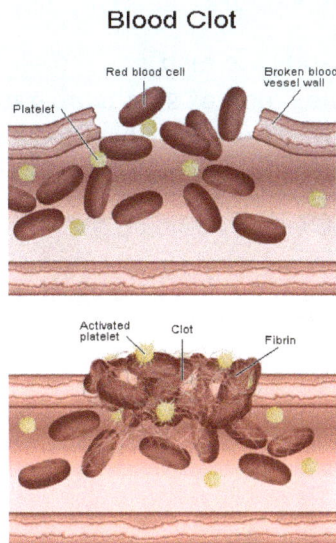

Blood Clot

Excessive blood loss can be fatal – this is why people with a condition known as haemophilia (low levels or absence of platelets) need medication otherwise even minor cuts can become fatal as bleeding continues without a scab being formed.

Alternatively, if platelet levels are excessively high then clotting within blood vessels can occur, leading to a stroke and or heart attack. This is why people with a history of cardiac problems are often prescribed medication to keep their blood thin to minimise the risk of clotting within their blood vessels.

Maintenance of Constant Body Temperature (Thermoregulation)

The core temperature range for a healthy adult is considered to be between 36.1°C and 37.8°C, with 37°C regarded as the average 'normal' temperature.

If the core temperature drops below this range it is known as hypothermia and if it rises above this range it is known as hyperthermia.

As temperatures move further into hypo or hyperthermia they become life threatening. Because of this the body works continuously to maintain its core temperature within the healthy range.

This process of temperature regulation in known as thermoregulation and the cardio-vascular system plays an integral part.

Temperature changes within the body are detected by sensory receptors called thermo-receptors, which in turn relay information about these changes to the hypothalamus in the brain.

When a deviation in temperature is recorded the hypothalamus reacts by initiating certain mechanisms in order to regain a safe temperature range. There are four sites where these adjustments in temperature can occur, they are:

a. Sweat glands: These glands are instructed to secrete sweat onto the surface of the skin when either the blood or skin temperature is detected to be above a normal safe temperature. This allows heat to be lost through evaporation and cools the skin so blood that has been sent to the skin can in turn be cooled.

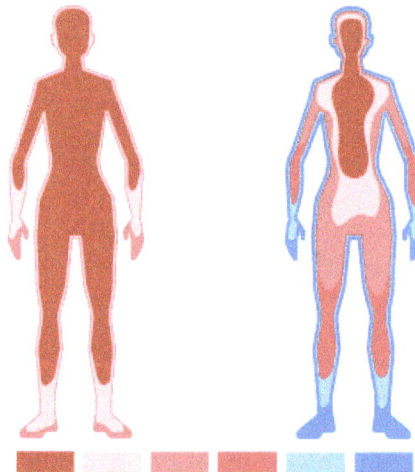

b. Smooth muscle around arterioles: Increases in temperature result in the smooth muscle in the walls of arterioles being stimulated to relax causing vasodilation (increase in diameter of the vessel).

This in turn increase the volume of blood flow to the skin, allowing cooling to occur. We see this is in the below diagram where blood that is normally concentrated around the core organs is shunted to the skin to cool when the body is under heat stress.

If however the thermoreceptors detect a cooling of the blood or skin then the hypothalamus reacts by sending a message to the smooth muscle of the arteriole walls causing the arterioles to vasoconstrict (reduce their diameter), thus reducing the blood flow to the skin and therefore helping to maintain core body temperature.

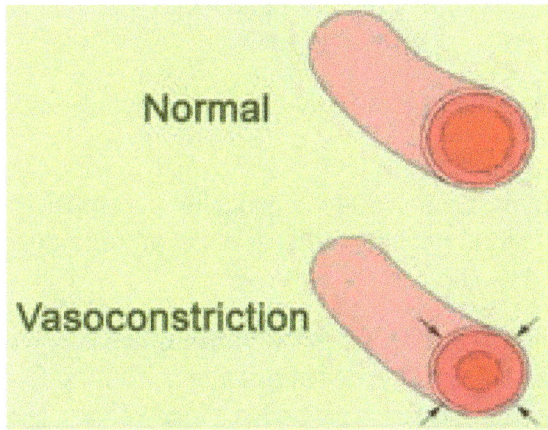

c. Skeletal muscle: When a drop in blood temperature is recorded the hypothalamus can also react by causing skeletal muscles to start shivering. Shivering is actually lots of very fast, small muscular contractions which produce heat to help warm the blood.

d. Endocrine glands: The hypothalamus may trigger the release of hormones such as thyroxin, ad renalin and noradrenalin in response to drops in blood temperature. These hormones all contribute to increasing the bodies metabolic rate (rate at which the body burns fuel) and therefore increasing the production of heat.

Maintaining Fluid Balance within the Body

The cardiovascular system works in conjunction with other body systems (nervous and endocrine) to balance the body's fluid levels. Fluid balance is essential in order to ensure sufficient and efficient movement of electrolytes, nutrients and gases through the body's cells.

When the fluid levels in the body do not balance a state of dehydration or hyperhydration can occur, both of which impede normal body function and if left unchecked can become dangerous or even fatal.

Dehydration is the excessive loss of body fluid, usually accompanied by an excessive loss of electrolytes.

The symptoms of dehydration include; headaches, cramps, dizziness, fainting and raised blood pressure (blood becomes thicker as its volume decreases requiring more force to pump it around the body).

Hyperhydration, on the other hand results from an excessive intake of water which pushes the normal balance of electrolytes outside of their safe limits. This can occur through long bouts of intensive exercise where electrolytes are not replenished and excessive amounts of water are consumed.

This can result in the recently consumed fluid rushing into the body's cells, causing tissues to swell. If this swelling occurs in the brain it can put excessive pressure on the brain stem that may result in seizures, brain damage, coma or even death.

Dehydration or a loss of body fluid (through sweat, urination, bleeding etc) results in an increase in 'blood tonicity' (the concentration of substances within the blood) and a decrease in blood volume. Where as hyperhydration or a gain in body fluid (intake of water) usually results in a reduction of blood tonicity and an increase in blood volume.

Any change in blood tonicity and volume is detected by the kidneys and osmoreceptors in the hypothalamus.

Osmoreceptors are specialist receptors that detect changes in the dilution of the blood. Essentially they detect if we are hydrated (diluted blood) or dehydrated (less diluted blood).

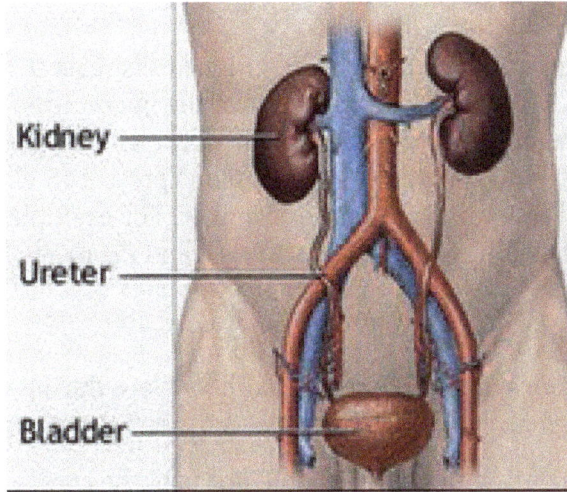

In response, hormones are released and transported by the cardiovascular system (through the blood) to act on target tissues such as the kidneys to increase or decrease urine production. Another way the cardiovascular system maintains fluid balance is by either dilating (widening) or constricting (tightening) blood vessels to increase or decrease the amount of fluid that can be lost through sweat.

Heart

The heart is a muscular organ about the size of a closed fist that functions as the body's circulatory pump. It takes in deoxygenated blood through the veins and delivers it to the lungs for oxygenation before pumping it into the various arteries (which provide oxygen and nutrients to body tissues by transporting the blood throughout the body). The heart is located in the thoracic cavity medial to the lungs and posterior to the sternum.

On its superior end, the base of the heart is attached to the aorta, pulmonary arteries and veins, and the vena cava. The inferior tip of the heart, known as the apex, rests just superior to the diaphragm. The base of the heart is located along the body's midline with the apex pointing toward the left side. Because the heart points to the left, about 2/3 of the heart's mass is found on the left side of the body and the other 1/3 is on the right.

Physiology of the Heart

Coronary Systole and Diastole

At any given time the chambers of the heart may found in one of two states:

- Systole: During systole, cardiac muscle tissue is contracting to push blood out of the chamber.

- Diastole: During diastole, the cardiac muscle cells relax to allow the chamber to fill with blood. Blood pressure increases in the major arteries during ventricular systole and decreases during ventricular diastole. This leads to the 2 numbers associated with blood pressure—systolic blood pressure is the higher number and diastolic blood pressure is the lower number. For example, a blood pressure of 120/80 describes the systolic pressure (120) and the diastolic pressure (80).

Cardiac Cycle

The cardiac cycle includes all of the events that take place during one heartbeat. There are 3 phases to the cardiac cycle: atrial systole, ventricular systole, and relaxation.

- Atrial systole: During the atrial systole phase of the cardiac cycle, the atria contract and push blood into the ventricles. To facilitate this filling, the AV valves stay open and the semilunar valves stay closed to keep arterial blood from re-entering the heart. The atria are much smaller than the ventricles, so they only fill about 25% of the ventricles during this phase. The ventricles remain in diastole during this phase.

- Ventricular systole: During ventricular systole, the ventricles contract to push blood into the aorta and pulmonary trunk. The pressure of the ventricles forces the semilunar valves to open and the AV valves to close. This arrangement of valves allows for blood flow from the ventricles into the arteries. The cardiac muscles of the atria repolarize and enter the state of diastole during this phase.

- Relaxation phase: During the relaxation phase, all 4 chambers of the heart are in diastole as blood pours into the heart from the veins. The ventricles fill to about 75% capacity during this phase and will be completely filled only after the atria enter systole. The cardiac muscle cells of the ventricles repolarize during this phase to prepare for the next round of depolarization and contraction. During this phase, the AV valves open to allow blood to flow freely into the ventricles while the semilunar valves close to prevent the regurgitation of blood from the great arteries into the ventricles.

Blood Flow through the Heart

Deoxygenated blood returning from the body first enters the heart from the superior and inferior vena cava. The blood enters the right atrium and is pumped through the tricuspid valve into the right ventricle. From the right ventricle, the blood is pumped through the pulmonary semilunar valve into the pulmonary trunk.

The pulmonary trunk carries blood to the lungs where it releases carbon dioxide and absorbs oxygen. The blood in the lungs returns to the heart through the pulmonary veins. From the pulmonary veins, blood enters the heart again in the left atrium.

The left atrium contracts to pump blood through the bicuspid (mitral) valve into the left ventricle. The left ventricle pumps blood through the aortic semilunar valve into the aorta. From the aorta, blood enters into systemic circulation throughout the body tissues until it returns to the heart via the vena cava and the cycle repeats.

Electrocardiogram

The electrocardiogram (also known as an EKG or ECG) is a non-invasive device that measures and monitors the electrical activity of the heart through the skin. The EKG produces a distinctive waveform in response to the electrical changes taking place within the heart.

The first part of the wave, called the P wave, is a small increase in voltage of about 0.1 mV that corresponds to the depolarization of the atria during atrial systole. The next part of the EKG wave is the QRS complex which features a small drop in voltage (Q) a large voltage peak (R) and another small drop in voltage (S). The QRS complex corresponds to the depolarization of the ventricles during ventricular systole. The atria also repolarize during the QRS complex, but have almost no effect on the EKG because they are so much smaller than the ventricles.

The final part of the EKG wave is the T wave, a small peak that follows the QRS complex. The T wave represents the ventricular repolarization during the relaxation phase of the cardiac cycle. Variations in the waveform and distance between the waves of the EKG can be used clinically to diagnose the effects of heart attacks, congenital heart problems, and electrolyte imbalances.

Heart Sounds

The sounds of a normal heartbeat are known as "lubb" and "dupp" and are caused by blood pushing on the valves of the heart. The "lubb" sound comes first in the heartbeat and is the longer of the two heart sounds. The "lubb" sound is produced by the closing of the AV valves at the beginning of ventricular systole. The shorter, sharper "dupp" sound is similarly caused by the closing of the semilunar valves at the end of ventricular systole. During a normal heartbeat, these sounds repeat in a regular pattern of lubb-dupp-pause. Any additional sounds such as liquid rushing or gurgling indicate a structure problem in the heart. The most likely causes of these extraneous sounds are defects in the atrial or ventricular septum or leakage in the valves.

Cardiac Output

Cardiac output (CO) is the volume of blood being pumped by the heart in one minute. The equation used to find cardiac output is:

CO = Stroke Volume x Heart Rate

Stroke volume is the amount of blood pumped into the aorta during each ventricular systole, usually measured in milliliters. Heart rate is the number of heartbeats per minute. The average heart can push around 5 to 5.5 liters per minute at rest.

Endocardium

The endocardium is the innermost layer of heart tissue that lines the cavities and valves of the heart. This layer is composed of loose connective tissue and simple squamous epithelial tissue. The endocardium regulates the contractions of the heart, aids cardiac development, and may regulate the composition of the blood that feeds the tissues of the heart.

The heart sits in a fluid-filled sac called the parietal pericardium. The tough, outer fibrous layer of the parietal pericardium protects the heart and roots it in place. The thin, inner serous layer connects the sac to the heart, which is composed of three layers. On the outside, the epicardium, also called the visceral pericardium, is composed of connective tissue and fat. The visceral pericardium connects loosely with the parietal pericardium and tightly with the myocardium, the middle layer of tissue in the heart.

The myocardium is composed of cardiac muscle and sits between the epicardium and endocardium. The myocardium is responsible for the contractions of the heart, which occur spontaneously, or without stimulation from the nervous system. These contractions allow blood to enter the atria and pump blood out of the ventricles. The endocardium is the inner layer of the heart that connects with the myocardium and lines the atria and ventricles.

Humans have four chambers in their hearts: the right ventricle and the left ventricle in the bottom two quadrants of the heart, and the right atrium and left atrium in the top two quadrants of the heart. The atria receive blood from the body and pass it on to the body through atrioventricular (AV) valves. The ventricles accept blood from the atria and pump it out into the body.

Blood that has already been circulated and "used" by the body is pumped into the right atrium, which then passes it on to the right ventricle. The right ventricle receives de-oxygenated blood from the right atrium and pumps the blood out to the lungs to pick up more oxygen. The left atrium takes back the re-oxygenated blood and passes it on to the left ventricle, which in turn pumps the blood into the body. The muscle in the myocardium executes the contractions that move the blood through the heart, control the valves between chambers, and pump blood out of the heart. The endocardium does not trigger these contractions, but it helps to regulate them.

The endocardium lines the walls of the atria and ventricles and the valves between them. The cellular make-up of the endocardium is close to that of the endothelium, the tissue layer that lines the inside of blood vessels. On its luminal side, or the side closest to the cavities of the heart, it is composed of simple squamous epithelium, a single layer of scaly cells. Underneath lies a layer of loose connective tissue, a tissue with variable, widely spaced fibers.

Usually, heart injuries resulting from heart attacks do not extend as far inward as the endocardium, but if they do, it can be very serious. Damage to the inner lining of the heart can negatively impact the heart's ability to contract at a quick, regular pace. Diseases such as endocarditis, a bacterial infection of the endocardium, are more typical in people with damaged heart valves.

Myocardium

The myocardium is the muscle layer of the heart, responsible for the heart's pumping action, which supplies the entire body with blood. The myocardium consists of cardiac muscle, a type of muscle unlike any other muscle in the body. Cardiac muscle combines features of skeletal muscle, which controls voluntary body movement, and smooth muscle, which controls the movement of all body organs other than the heart.

The myocardium is the middle layer of the cardiac wall; the outermost layer is the epicardium, while the innermost is the endocardium. The epicardium consists mostly of connective tissue and serves to protect the inner structures of the heart. The endocardium is a thin layer of epithelial cells, similar to that which lines the inside of blood vessels.

The cardiac muscle that makes up this structure is involuntary, like the smooth muscle in the body's other organs. Involuntary muscle is not under conscious control, and contrasts with voluntary skeletal muscle, which is attached to the skeleton and used for skeletal movement like walking and standing. Cardiac muscle is more similar in structure, however, to skeletal muscle than to smooth muscle. Both cardiac muscle and skeletal muscle are striated, meaning the muscle fibers are arranged into parallel bundles, and have alternating thick and thin protein filaments. Striated muscle is better suited to brief, intense contractions than smooth muscle.

While skeletal muscle fibers are arranged into regular, non-branching bundles, the muscle fibers of the myocardium branch at irregular angles, and connect to other muscle cells at junctions called intercalated discs. The cells that make up cardiac muscle are called cardiac myocytes, or cardiomyocytes. They also differ from skeletal muscle cells in that they require extracellular calcium for contraction to take place.

The contractions of the myocardium are responsible for pumping oxygenated blood throughout the body, providing the body with the oxygen and other nutrients it needs to function properly. The heart muscle also pumps deoxygenated blood into the lungs so that it can be oxygenated again. After the blood has delivered oxygen throughout the body, deoxygenated blood returns to the heart, which in turn pumps the blood into the lungs. After the blood is reoxygenated in the lungs, it returns to the heart to be pumped throughout the body once again. Like all body tissues, the myocardium itself requires a blood supply in order to function; the coronary arteries supply the heart muscle with blood.

Epicardium

The epicardium is a layer of muscle located on the outside of the heart. A continuous piece of muscle, this tissue performs a protective role, helping hold the other muscles close together. The heart is responsible for pumping blood through the blood vessels, causing it to circulate throughout the body.

The heart is comprised completely of cardiac muscle and is found in all vertebrates. The heart must be functioning for the animal to live. The epicardium is considered to be an involuntary muscle. This means the muscle contracts and releases based on automated signals from the brain.

Several mechanical devices have been invented to mimic the function of the heart muscles. These artificial heart machines are used to keep the blood oxygenated and moving through the body during heart surgery, or in situations where the person has suffered catastrophic injuries. The machine requires significant power and puts a great deal of strain on the patient's body. As a result, it is only used in emergency situations.

The epicardium cannot be separated from the other heart muscles, but it can be stimulated by electronic pulses to change the rhythm of the heart. This type of device is usually implanted into the patient, and is used when the heart is unable to sustain a consistent rhythm or is prone to skipping a beat. The use of implants and other devices to sustain life grew significantly after success with pacemakers, which are used to control the muscular contractions of the epicardium muscles.

The most serious risk to the heart muscle is obesity. The buildup of fatty deposits on the heart muscles contributes directly to heart attacks, which are one of the leading causes of natural death among adults. A diet rich is fruits, vegetables, and whole grains significantly reduces the buildup of fat in the heart. Add a regular exercise routine to keep the heart muscle strong, and the risk is even lower.

Replacing the entire heart muscle is called a heart transplant, and is a very dangerous surgery, performed only when absolutely necessary. The heart must be donated from someone of similar blood type, height and weight who has passed away from a non-heart-related trauma. The list of people in need of a heart transplant exceeds the number of hearts available. People often pass away from heart disease or related complications while waiting for a suitable heart to become available.

Pericardium

The pericardium is a thin sac that surrounds your heart. It protects and lubricates your heart and keeps it in place within your chest.

Problems can occur when the pericardium becomes enflamed or fills with fluid. The swelling can damage your heart and affect its function.

Work of Pericardium

The pericardium has a few important roles:

- It keeps your heart fixed in place within your chest cavity.

- It prevents your heart from stretching too much and overfilling with blood.

- It lubricates your heart to prevent friction with the tissues around it as it beats.

- It protects your heart from any infections that might spread from nearby organs like the lungs.

Pericardium Layers

The pericardium has two layers:

- Fibrous pericardium is the outer layer. It's made from thick connective tissue and is attached to your diaphragm. It holds your heart in place in the chest cavity and protects from infections.

- Serous pericardium is the inner layer. It's further divided into two more layers: the visceral and parietal layers. The serous pericardium helps to lubricate your heart.

In between these two layers is the fluid-filled pericardial cavity. It lubricates the heart and protects it from injury.

Pericardial Effusion

Pericardial effusion is the buildup of too much fluid between the pericardium and your heart. This can happen from damage or disease in the pericardium. Fluid can also build up if there's bleeding in your pericardium after an injury.

Possible causes of pericardial effusion include:

- Diseases that cause inflammation, such as lupus or rheumatoid arthritis,

- Severe underactive thyroid (hypothyroidism),

- Infections,

- Recent heart surgery,

- Cancer that has spread to your pericardium,

- Kidney failure.

Symptoms of pericardial effusion include:

- Chest pressure or pain,

- Shortness of breath,

- Difficulty breathing when you lie down,

- Nausea,

- A feeling of fullness in your chest,

- Trouble swallowing.

The excess fluid from pericardial effusion can cause intense pressure on your heart and damage it.

Pericardial Cyst

A pericardial cyst is a noncancerous, fluid-filled growth in the pericardium. This type of cyst is very rare, affecting only 1 in 100,000 people.

Most people who have pericardial cysts are born with them, but they often aren't diagnosed until they reach their 20s or 30s.

Pericardial cysts are usually found during a chest X-ray that's done for another reason since these cysts don't cause symptoms on their own.

Symptoms may only appear when the cyst presses on nearby organs or structures, and can include:

- Pain in your right shoulder that radiates to your left shoulder,

- Shortness of breath,

- Rapid, strong heart rate (palpitations),

- A feeling of fullness in your chest.

Pericardial cysts aren't dangerous themselves. However, if they press on your lungs or other structures in your chest, they can cause complications like inflammation or severe bleeding. Rarely, a pericardial cyst can lead to heart failure.

Other Problems with the Pericardium

A few other conditions and complications can also affect the pericardium.

Pericarditis

Pericarditis is swelling of the pericardium. Possible causes include:

- Infection with a virus, bacteria, or fungus,

- Autoimmune disorders such as lupus, rheumatoid arthritis, and scleroderma,

- Heart attack,

- Heart surgery,

- Injuries, such as from a car accident,

- Kidney failure,

- Tuberculosis,

- Medications such as phenytoin (Dilantin), warfarin (Coumadin), and procainamide.

Acute pericarditis starts suddenly and lasts only a few weeks. Chronic pericarditis develops more slowly and can last longer.

Usually pericarditis is mild and heals over time. Sometimes it will improve with plenty of rest. More severe pericarditis may need to be treated with medication or surgery to prevent it from damaging your heart.

Cardiac Tamponade

Cardiac tamponade is a condition that's caused by a buildup of fluid, blood, gas, or a tumor in your pericardial cavity. This buildup places pressure on your heart, which prevents it from filling and emptying properly.

Cardiac tamponade is not the same as pericardial effusion, though it can be a complication of fluid buildup from pericardial effusion.

One sign of cardiac tamponade is a large drop in blood pressure. Cardiac tamponade is a medical emergency. It can be life threatening if it isn't treated quickly.

Arteries

The arteries are the blood vessels that deliver oxygen-rich blood from the heart to the tissues of the body. Each artery is a muscular tube lined by smooth tissue and has three layers:

- The intima, the inner layer lined by a smooth tissue called endothelium;

- The media, a layer of muscle that lets arteries handle the high pressures from the heart;

- The adventitia, connective tissue anchoring arteries to nearby tissues.

The largest artery is the aorta, the main high-pressure pipeline connected to the heart's left ventricle. The aorta branches into a network of smaller arteries that extend throughout the body. The arteries' smaller branches are called arterioles and capillaries. The pulmonary arteries carry oxygen-poor blood from the heart to the lungs under low pressure, making these arteries unique.

Types of Arteries

Cartoid Artery

The carotid artery is a major artery of the head and neck. There are two carotid arteries, one on the left and one on the right. From their origins and for about half their length, the carotid arteries are known as common carotid arteries. The left carotid arises from the arch of the aorta, while the right carotid arises as one of the branches of the bifurcation of the brachiocephalic artery (trunk) into the carotid and right subclavian artery. The carotids then continue along similar paths within their respective sides of the neck and skull. At approximately the level of the third cervical vertebra, the common carotid branches into the internal and external carotid arteries.

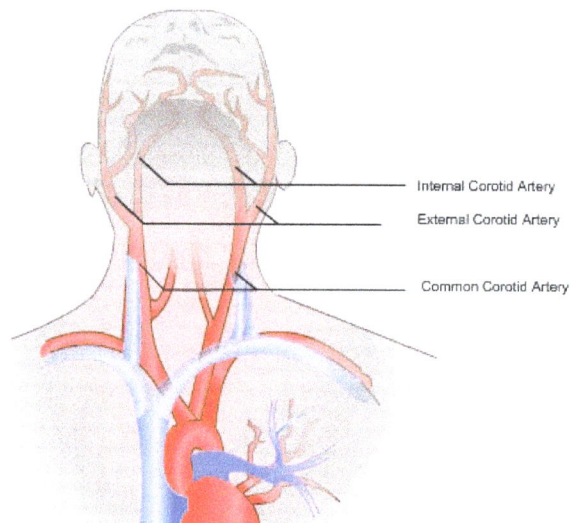

Internal Corotid Artery

External Corotid Artery

Common Corotid Artery

Iliac Artery

The Iliac artery is one of the large arteries supplying blood to the pelvis and legs. The iliac artery originates from the common iliac artery, with branches to the inferior epigastric and deep circumflex iliac arteries, becoming the femoral artery at the inguinal ligament; external iliac artery.

Femoral Artery

The femoral artery is a large artery of the thigh. It is a continuation of the external iliac artery which comes from the abdominal aorta. The external iliac artery becomes known as the femoral artery after it passes the inguinal ligament. For a while at this location, (the femoral triangle), it can be known as the common femoral, because it has not yet branched. It usually gives off a branch known as the profunda femoris or the deep artery of the thigh , while continuing down the thigh medial to the femur. (The profunda femoris is even closer to the femur, and is more posterior).

The femoral artery goes through the adductor hiatus (a hole in the tendon of adductor magnus), into the posterior of the knee. Passing between the condyles of the femur, it becomes the popliteal artery of the popliteal fossa.

Radial Artery

The radial artery is the main blood vessel, with oxygenated blood, of the lateral aspect of the forearm. It arises from the brachial artery and terminates in the deep palmar arch, which joins with the deep branch of the ulnar artery. It is palpable on the anterior aspect of the arm over the carpal bones (where it is commonly used to assess the heart rate and cardiac rhythm).

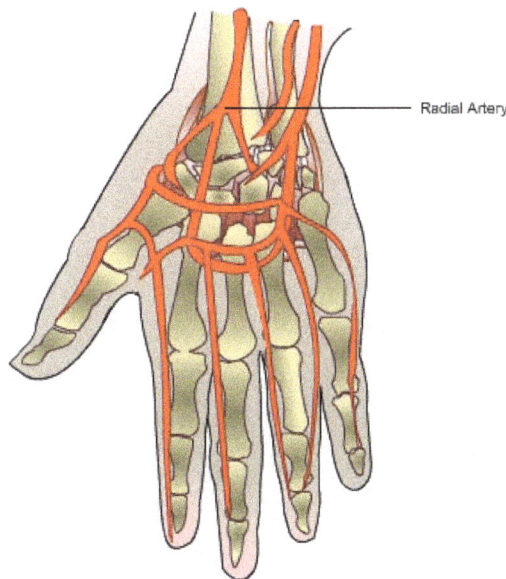

Radial Artery

Structure of Arteries

Arteries contain a high percentage of a special type of muscle, called smooth muscle, that can be controlled by hormones and special signals from the nervous system. The outer layer of an artery is made of collagen fibers. The middle layer has smooth muscle and elastic fibers. The inner layer is the lining called the endothelium.

Blood travels through the hollow center of the arteries. If this hollow center becomes constricted due to overdevelopment of the muscle or the formation of plaques, it can raise blood pressure. Plaque also makes the arteries less flexible. If an artery ruptures or is blocked, such as in a stroke or heart attack, the tissues that it normally supplies will die.

The thick, strong walls of arteries make them able to resist the high pressures that exist near the heart. All of the major organs in the body have their own special kind of arteries which are uniquely structured to deliver the supplies needed.

The heart muscle is supplied by the coronary arteries. The left coronary artery and the right coronary artery branch off of the aorta and the left coronary artery further divides into the circumflex artery and the left anterior descending artery. These four arteries are the ones that may be replaced in coronary artery bypass graft (CABG) surgery. A quadruple bypass replaces all four arteries.

Capillaries

Capillaries are the smallest blood vessels in the body, connecting the smallest arteries to the smallest veins. These vessels are often referred to as the "microcirculation."

Structure of Capillaries

Capillaries are very thin, approximately 5 micrometers in diameter, and are composed of only two layers of cells; an inner layer of endothelial cells and an outer layer of epithelial cells. They are so small that red blood cells need to flow through them single file. If all the capillaries in the human body were lined up in single file, the line would stretch over 100,000 miles. It's been estimated that there are 40 billion capillaries in the average human body. Surrounding this layer of cells is something called the basement membrane, a layer of protein surrounding the capillary.

Capillaries in the Circulatory System

Capillaries may be thought of as the central portion of circulation. Blood leaves the heart through the aorta and the pulmonary arteries traveling to the rest of the body and to the lungs respectively. These large arteries become smaller arterioles and eventually narrow to form the capillary bed. From the capillaries, blood flows into the smaller venules and then into veins, flowing back to the heart.

Function of Capillaries

The capillaries are responsible for facilitating the transport and exchange of gases, fluids, and nutrients in the body.

Gas Exchange

In the lungs, oxygen diffuses from the alveoli into capillaries to be attached to hemoglobin and be carried throughout the body. Carbon dioxide (from deoxygenated blood) in turn flows from the capillaries back into alveoli to be exhaled into the environment.

Fluid and Nutrient Exchange

Likewise, fluids and nutrients diffuse through selectively permeable capillaries into the tissues of the body, and waste products are picked up in the capillaries to be transported through veins to the kidneys and liver where they are thus processed and eliminated from the body.

Types of Capillaries

There are 3 primary types of capillaries:

- Continuous - These capillaries have no perforations and allow only small molecular to pass through. They are present in muscle, skin, fat, and nerve tissue.

- Fenestrated - These capillaries have small pores which allow small molecules through and are located in the intestines, kidneys, and endocrine glands.

- Sinusoidal or discontinuous - These capillaries have large open pores—large enough to allow a blood cell through. They are present in the bone marrow, lymph nodes, and the spleen, and are, in essence, the "leakiest" of the capillaries.

- Blood-brain barrier - In the central nervous system the capillaries make up what is known as the blood-brain barrier. This barrier limits the ability of toxins (and, unfortunately, many chemotherapy agents) to pass through into the brain.

Blood Flow through Capillaries

Since the blood flow through capillaries plays such an important part in maintaining the body, you may wonder what happens when blood flow changes, for example, if your blood pressure would drop (hypotension.) Capillary beds are regulated through something called autoregulation, so that if blood pressure would drop, flow through the capillaries will continue to provide oxygen and nutrients to the tissues of the body. With exercise, more capillary beds are recruited in the lungs to prepare for an increased need for oxygen in tissues of the body.

The flow of blood in the capillaries is controlled by precapillary sphincters. A precapillary sphincter is the muscular fibers that control the movement of blood between the arterioles and capillaries.

Capillary Microcirculation

Regulation of fluid movement between the capillaries and the surrounding interstitial tissues is determined by the balance of two forces: the hydrostatic pressure and osmotic pressure.

On the arterial side of the capillary, the hydrostatic pressure (the pressure that comes from the heart pumping blood and the elasticity of the arteries) is high. Since capillaries are "leaky" this pressure forces fluid and nutrients against the walls of the capillary and out into the interstitial space and tissues.

On the vein side of the capillary, the hydrostatic pressure has dropped significantly. At this point, it is the osmotic pressure of the fluid within the capillary (due to the presence of salts and proteins in the blood) that draws fluids back into the capillary. Osmotic pressure is also referred to as oncotic pressure and is what pulls fluids and waste products out of the tissues and into the capillary to be returned to the bloodstream (and then delivered to the kidneys among other sites.)

Number of Capillaries

The number of capillaries in a tissue can vary widely. Certainly, the lungs are packed with capillaries surrounding the alveoli to pick up oxygen and drop off carbon dioxide. Outside of the lungs, capillaries are more abundant in tissues that are more metabolically active.

Capillaries 'Visually'

Skin blanching - If you've ever wondered why your skin turns white when you put pressure on it the answer is the capillaries. Pressure on the skin presses blood out of the capillaries resulting in the blanching or pale appearance when the pressure is removed.

Capillary refill - Doctors often check for "capillary refill." This is tested by observing how rapidly the skin becomes pink again after pressure is released and can give an idea of the health of the tissues. An example of this use would be in people with burns. A second-degree burn may reveal capillary refill to be somewhat delayed, but in a third-degree burn, there would be no capillary refill at all.

Emergency responders often check capillary refill by pushing on a fingernail or toenail, then releasing pressure and waiting to see how long it takes for the nailbed to appear pink again. If color returns within two seconds (the amount of time it takes to say capillary refill), circulation to the arm or leg is probably okay. If capillary refill takes more than two seconds, the circulation of the limb is probably compromised and considered an emergency. There are other settings in which capillary refill is delayed as well, such as in dehydration.

Third spacing and capillary permeability - You may hear doctors talk about a phenomenon known as "third spacing." Capillary permeability refers to the ability of fluids to pass out of the capillaries into the surrounding tissues. Capillary permeability can be increased by cytokines (leukotrienes, histamines, and prostaglandins) released by cells of the immune system. The increased fluid (third spacing) locally can result in hives. When someone is very ill, this third spacing due to leaky capillaries may be widespread, giving their body a swollen appearance.

Capillary blood samples - Most of the time when you have your blood drawn, a technician will take blood from a vein in your arm. Capillary blood may also be used to do some blood tests, such as for those who monitor their blood sugar. A lancet is used to cut the finger (cut capillaries) and can be used for testing blood sugar and blood pH.

Conditions Involving the Capillaries

There are several both common and uncommon conditions which involve the capillaries. A few of these include:

Port wine stain - "birth mark":- Around 1 in 300 children are born with "birth marks" consisting of an area of red or purple skin related to dilated capillaries. Most port wine stains are a cosmetic problem rather than a medical concern, but they may bleed easily when irritated.

Capillary malformation - arteriovenous malformation syndrome: Capillary malformation may occur as part of an inherited syndrome present in roughly 1 in 100,000 people of European ancestry. In this syndrome, there is more blood flow than normal through the capillaries near the skin, which results in pink and red dots on the skin. The may occur alone, or people may have other complications of this syndrome such as arteriovenous malformations (abnormal connections between arteries and veins) which, when in the brain, can cause headaches and seizures.

Systemic capillary leak syndrome : A rare disorder known as capillary leak syndrome involves leaky capillaries which result in constant nasal congestion and episodes of fainting due to rapid drops in blood pressure.

Macular degeneration: Macular degeneration, now the leading cause of blindness in the United States, occurs secondary to damage in the capillaries of the retina.

Veins

Veins are a type of blood vessel that return deoxygenated blood from your organs back to your heart. These are different from your arteries, which deliver oxygenated blood from your heart to the rest of your body.

Deoxygenated blood that flows into your veins is collected within tiny blood vessels called capillaries. Capillaries are the smallest blood vessels in your body. Oxygen passes through the walls of your capillaries to your tissues. Carbon dioxide can also move into your capillaries from the tissue before entering your veins.

The venous system refers to the network of veins that work to deliver deoxygenated blood back to your heart.

Vein Structure

The walls of your veins are made up of three different layers:

- Tunica externa: This is the outer layer of the vein wall, and it's also the thickest. It's mostly made up of connective tissue. The tunica externa also contains tiny blood vessels called vasa vasorum that supply blood to the walls of your veins.

- Tunica media: The tunica media is the middle layer. It's thin and contains a large amount of collagen. Collagen is one of the main components of connective tissue.

- Tunica intima: This is the innermost layer. It's a single layer of endothelium cells and some connective tissue. This layer sometimes contains one-way valves, especially in the veins of your arms and legs. These valves prevent blood from flowing backward.

Types of Veins

Veins are often categorized based on their location and any unique features or functions.

Pulmonary and Systemic Veins

Your body circulates blood on two different tracks called the systemic circuit and the pulmonary circuit. Veins are based on the circuit they're found in:

- Pulmonary veins: The pulmonary circuit carries deoxygenated blood from your heart to your lungs. Once your lungs oxygenate the blood, the pulmonary circuit brings it back to your heart. There are four pulmonary veins. They're unique because they carry oxygenated blood. All other veins carry only deoxygenated blood.

- Systemic veins: The systemic circuit carries deoxygenated blood from the rest of the body back to your heart, where it then enters the pulmonary circuit for oxygen. Most veins are systemic veins.

Deep Veins and Superficial Veins

Systemic veins are further classified as being either:

- Deep veins: These are found in muscles or along bones. The tunica intima of a deep vein usually has a one-way valve to prevent blood from flowing backward. Nearby muscles also compress the deep vein to keep blood moving forward.

- Superficial veins: These are located in the fatty layer under your skin. The tunica intima of a superficial vein can also have a one-way valve. However, without a nearby muscle for compression, they tend to move blood more slowly than deep veins do.

- Connecting veins: Blood from superficial veins is often directed into the deep veins through short veins called connecting veins. Valves in these veins allow blood to flow from the superficial veins to your deep veins, but not the other way.

Conditions that Affect the Venous System

Many conditions can affect your venous system. Some of the most common ones include:

- Deep Vein Thrombosis (DVT): A blood clot forms in a deep vein, usually in your leg. This clot can potentially travel to your lungs, causing pulmonary embolism.

- Superficial Thrombophlebitis: An inflamed superficial vein, usually in your leg, develops a blood clot. While the clot can occasionally travel to a deep vein, causing DVT, thrombophlebitis is generally less serious than DVT.

- Varicose Veins: Superficial veins near the surface of the skin visibly swell. This happenswhen one-way valves break down or vein walls weaken, allowing blood to flow backward.

- Chronic Venous Insufficiency: Blood collects in the superficial and deep veins of your legs due to improper functioning of one-way valves. While similar to varicose veins, chronic venous insufficiency usually causes more symptoms, including coarse skin texture and ulcers in some cases.

Symptoms of a Venous Condition

While the symptoms of a venous condition can vary widely, some common ones include:

- Inflammation or swelling,

- Tenderness or pain,

- Veins that feel warm to the touch,

- A burning or itching sensation.

These symptoms are especially common in your legs. If you notice any of these and they don't improve after a few days, make an appointment with your doctor.

They can perform a venography. In this procedure, your doctor injects contrast die into your veins to produce an X-ray image of a particular area.

Tips for Healthy Veins

Follow these tips to keep your vein walls and valves strong and properly functioning:

- Get regular exercise to keep blood moving through your veins.

- Try to maintain a healthy weight, which reduces your risk of high blood pressure. High blood pressure can weaken your veins overtime due to added pressure.

- Avoid long periods of standing or sitting. Try to change positions regularly throughout the day.

- When sitting down, avoid crossing your legs for long periods of time or regularly switch positions so one leg isn't on top for a long period of time.

- When flying, drink plenty of water and try to stand up and stretch as often as possible. Even while sitting, you can flex your ankles to encourage blood flow.

Hepatic Portal Vein

The hepatic portal vein is a vessel that moves blood from the spleen and gastrointestinal tract to the liver.

It is approximately three to four inches in length and is usually formed by the merging of the superior mesenteric and splenic veins behind the upper edge of the head of the pancreas. In some individuals, the inferior mesenteric vein may enter this intersection instead.

In most people, the portal vein splits into left and right veins before entering the liver. The right vein then branches off into anterior and superior veins.

The portal vein supplies approximately 75 percent of blood flow to the liver. The portal vein is not a true vein, which means it does not drain into the heart. Instead, it brings nutrient-rich blood to the liver from the gastrointestinal tract and spleen. Once there, the liver can process the nutrients from the blood and filter out any toxic substances it contains before the blood goes back into general circulation.

Abnormally high blood pressure in the portal vein is known as portal hypertension. The condition may cause the growth of new blood vessels that bypass the liver, which can result in the circulation of unfiltered blood throughout the body. Portal hypertension is one of the potential serious complications of liver cirrhosis, which is a condition where normal liver tissue is replaced with scar tissue.

Cardiac Cycle

The cardiac cycle is the series of electrical impulses and muscle contractions that pressurizes different chambers of the heart, causing blood to flood in one direction. The cardiac cycle varies in different organisms, due to changes in the structure of the heart. Some organisms have a three-chambered heart, which consists of the sinus venosus, atrium, and ventricle. Most tetrapods have developed a more efficient heart, which can supply a greater pressure of blood to the organisms. The "four-chambered heart", is actually just a modification of the three chambered heart. The sinus venosus is reduced to the sinoatrial node, located on the right atrium. The atrium and ventricles are divided in the four-chambered heart, allowing a separate pathway to be established to the lungs, allowing for greater oxygenation of the blood. The separate circulations pathways are known as pulmonary (lungs) and systemic (body) circulation.

The following cardiac cycle phases are representative of the mammalian, four-chambered heart. The cardiac cycle of animals with three-chambered hearts is similar, except the atria and ventricles are not divided completely, if at all. In most animals, the heartbeat is regulated by nerves in the sinoatrial node, and carried out by nerves throughout the heart. Heart muscles cells are also connected laterally, allowing them to pass the nerve impulse received to all their neighbors, creating rhythmic contractions. Hagfish, and other organisms that have a more ancestral heart, simply use the heart to move liquid through their body at a slow rate. In organisms such as this, the cardiac cycle is much less distinguished because the heart does not set up a specific rhythm.

Cardiac Cycle Phases

The two main phases of the cardiac cycle are systole and diastole, and follow each other in sequence. Typically, the cardiac cycle starts with diastole, or the relaxation of all the heart muscles. During diastole, blood returns to the heart, and begins to fill the atria and ventricles. The lack of pressure in the ventricle allows the mitral and tricuspid valves to open, which allow blood from the atria into the left and right ventricles, respectively. This phase of the cardiac cycle can be seen in the image.

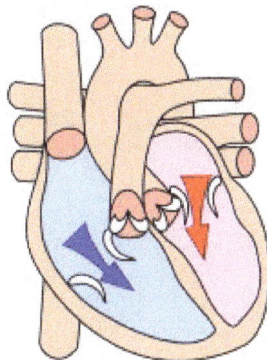

A signal sent to the sinoatrial node induces the muscles of both atria to contract. In unison, this forces blood out of the atria and into the ventricles. Most of the blood leaves the atria at this point in the cardiac cycle. As the atria squeeze, the action potential is passed through the muscles and nerves of the heart to the ventricles. Another wave of contraction starts, and the ventricles enter ventricular systole, and begin contracting themselves. The increased pressure in the ventricles closes the mitral and tricuspid valves, and opens the aortic and pulmonary valves. This can be seen in the image.

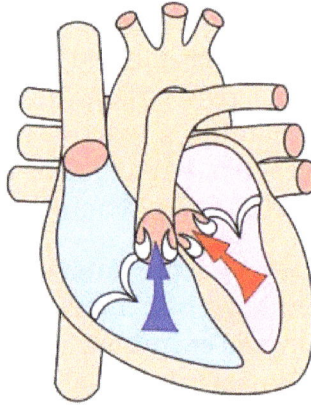

The ventricles contract hard, pushing most of blood volume they contain into the pulmonary and systemic circulation. The aorta is the main artery that feeds oxygenated blood to the body, and is attached to the left ventricle. The pulmonary artery exits the right ventricle and carries unoxygenated blood to the lungs. This blood then returns and enters the left atrium, to be pumped out to the body. The body used the oxygen and return the blood to the right atrium, for the cycle to start over. While it may take a while for one pump of blood to circle the body, the heart continues the cardiac cycle indefinitely, to ensure the movement of nutrients and oxygen in the body, as well as remove toxic metabolic wastes. The entire cardiac cycle can be seen in the following image, which tracts the cardiac cycle along with the pressure and volume of different chambers.

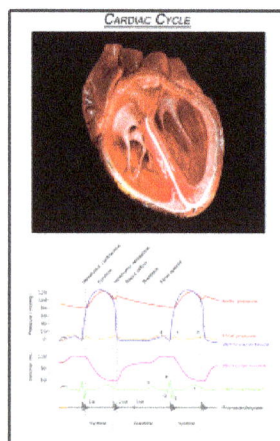

An import tool to measure the cardiac cycle is the electrocardiogram, which can be seen as the green line on the above animated graph. The electrocardiogram is a signal which can be measured by sensitive medical electronics, and provides a glimpse of the cardiac cycle, and the actions taking place in the heart. The "QRS" on the green line indicates significant points in the signal, and correspond to the contraction of the ventricles. The study of electrocardiograms can lead to important insights into the functioning of the heart, and significance is put only on the presence of a signal, but the strength of the signal and the spacing of the events.

References

- Cardiovascular-system-structure-function: schoolworkhelper.net, Retrieved 10 June 2018
- Major-functions-of-the-cardiovascular-system, anatomy-and-physiology: ptdirect.com, Retrieved
- 30 May 2018
- What-is-the-endocardium: wisegeek.com, Retrieved 15 April 2018
- What-is-the-myocardium: wisegeek.com, Retrieved 25 June 2018
- What-are-capillaries-2249069: verywellhealth.com, Retrieved 14 March 2018
- Portal-vein, human-body-maps: healthline.com, Retrieved 27 June 2018

7

Respiratory System

The respiratory system is a biological system that consists of organs and structures for the exchange of gases in humans. The topics elaborated in this chapter on upper and lower respiratory system will help in providing a better perspective about the human respiratory system.

The cells of the human body require a constant stream of oxygen to stay alive. The respiratory system provides oxygen to the body's cells while removing carbon dioxide, a waste product that can be lethal if allowed to accumulate. There are 3 major parts of the respiratory system: the airway and the muscles of respiration. The airway, which includes the nose, mouth, pharynx, larynx, trachea, bronchi, and bronchioles, carries air between the lungs and the body's exterior.

The Lung act as the functional units of the respiratory system by passing oxygen into the body and carbon dioxide out of the body. Finally, the muscles of respiration, including the diaphragm and intercostal muscles, work together to act as a pump, pushing air into and out of the lungs during breathing.

Anatomy of the Respiratory System

Nose and Nasal Cavity

The nose and nasal cavity form the main external opening for the respiratory system and are the first section of the body's airway—the respiratory tract through which air moves. The nose is a structure of the face made of cartilage, bone, muscle, and skin that supports and protects the anterior portion of the nasal cavity. The nasal cavity is a hollow space within the nose and skull that is lined with hairs and mucus membrane. The function of the nasal cavity is to warm, moisturize, and filter air entering the body before it reaches the lungs. Hairs and mucus lining the nasal cavity help to trap dust, mold, pollen and other environmental contaminants before they can reach the inner portions of the body. Air exiting the body through the nose returns moisture and heat to the nasal cavity before being exhaled into the environment.

Mouth

The mouth, also known as the oral cavity, is the secondary external opening for the respiratory tract. Most normal breathing takes place through the nasal cavity, but the oral cavity can be used to supplement or replace the nasal cavity's functions when needed. Because the pathway of air entering the body from the mouth is shorter than the pathway for air entering from the nose, the mouth does not warm and moisturize the air entering the lungs as well as the nose performs this function. The mouth also lacks the hairs and sticky mucus that filter air passing through the nasal cavity. The one advantage of breathing through the mouth is that its shorter distance and larger diameter allows more air to quickly enter the body.

Pharynx

The pharynx, also known as the throat, is a muscular funnel that extends from the posterior end of the nasal cavity to the superior end of the esophagus and larynx. The pharynx is divided into 3 regions: the nasopharynx, oropharynx, and laryngopharynx. The nasopharynx is the superior region of the pharynx found in the posterior of the nasal cavity. Inhaled air from the nasal cavity passes into the nasopharynx and descends through the oropharynx, located in the posterior of the oral cavity. Air inhaled through the oral cavity enters the pharynx at the oropharynx. The inhaled air then descends into the laryngopharynx, where it is diverted into the opening of the larynx by the epiglottis. The epiglottis is a flap of elastic cartilage that acts as a switch between the trachea and the esophagus. Because the pharynx is also used to swallow food, the epiglottis ensures that air passes into the trachea by covering the opening to the esophagus. During the process of swallowing, the epiglottis moves to cover the trachea to ensure that food enters the esophagus and to prevent choking.

Larynx

The larynx, also known as the voice box, is a short section of the airway that connects the laryngopharynx and the trachea. The larynx is located in the anterior portion of the neck, just inferior to the hyoid bone and superior to the trachea. Several cartilage structures make up the larynx and give it its structure. The epiglottis is one of the cartilage pieces of the larynx and serves as the cover of the larynx during swallowing. Inferior to the epiglottis is the thyroid cartilage, which is often referred to as the Adam's apple as it is most commonly enlarged and visible in adult males. The thyroid holds open the anterior end of the larynx and protects the vocal folds. Inferior to the thyroid cartilage is the ring-shaped cricoid cartilage which holds the larynx open and supports its posterior end. In addition to cartilage, the larynx contains special structures known as vocal folds, which allow the body to produce the sounds of speech and singing. The vocal folds are folds of mucous membrane that vibrate to produce vocal sounds. The tension and vibration speed of the vocal folds can be changed to change the pitch that they produce.

Trachea

The trachea, or windpipe, is a 5-inch long tube made of C-shaped hyaline cartilage rings lined with pseudostratified ciliated columnar epithelium. The trachea connects the larynx to the bronchi and allows air to pass through the neck and into the thorax.

The rings of cartilage making up the trachea allow it to remain open to air at all times. The open end of the cartilage rings faces posteriorly toward the esophagus, allowing the esophagus to expand into the space occupied by the trachea to accommodate masses of food moving through the esophagus.

The main function of the trachea is to provide a clear airway for air to enter and exit the lungs. In addition, the epithelium lining the trachea produces mucus that traps dust and other contaminants and prevents it from reaching the lungs. Cilia on the surface of the epithelial cells move the mucus superiorly toward the pharynx where it can be swallowed and digested in the gastrointestinal tract.

Bronchi and Bronchioles

At the inferior end of the trachea, the airway splits into left and right branches known as the primary bronchi. The left and right bronchi run into each lung before branching off into smaller secondary bronchi. The secondary bronchi carry air into the lobes of the lungs—2 in the left lung and 3 in the right lung. The secondary bronchi in turn split into many smaller tertiary bronchi within each lobe. The tertiary bronchi split into many smaller bronchioles that spread throughout the lungs. Each bronchiole further splits into many smaller branches less than a millimeter in diameter called terminal bronchioles. Finally, the millions of tiny terminal bronchioles conduct air to the alveoli of the lungs.

As the airway splits into the tree-like branches of the bronchi and bronchioles, the structure of the walls of the airway begins to change. The primary bronchi contain many C-shaped cartilage rings that firmly hold the airway open and give the bronchi a cross-sectional shape like a flattened circle or a letter D. As the bronchi branch into secondary and tertiary bronchi, the cartilage becomes more widely spaced and more smooth muscle and elastin protein is found in the walls. The bronchioles differ from the structure of the bronchi in that they do not contain any cartilage at all. The presence of smooth muscles and elastin allow the smaller bronchi and bronchioles to be more flexible and contractile.

The main function of the bronchi and bronchioles is to carry air from the trachea into the lungs. Smooth muscle tissue in their walls helps to regulate airflow into the lungs. When greater volumes of air are required by the body, such as during exercise, the smooth muscle relaxes to dilate the bronchi and bronchioles. The dilated airway provides less resistance to airflow and allows more air to pass into and out of the lungs. The smooth muscle fibers are able to contract during rest to prevent hyperventilation. The bronchi and bronchioles also use the mucus and cilia of their epithelial lining to trap and move dust and other contaminants away from the lungs.

Lungs

The lungs are a pair of large, spongy organs found in the thorax lateral to the heart and superior to the diaphragm. Each lung is surrounded by a pleural membrane that provides the lung with space to expand as well as a negative pressure space relative to the body's exterior. The negative pressure allows the lungs to passively fill with air as they relax. The left and right lungs are slightly different in size and shape due to the heart pointing to the left side of the body. The left lung is therefore slightly smaller than the right lung and is made up of 2 lobes while the right lung has 3 lobes.

The interior of the lungs is made up of spongy tissues containing many capillaries and around 30 million tiny sacs known as alveoli. The alveoli are cup-shaped structures found at the end of the terminal bronchioles and surrounded by capillaries. The alveoli are lined with thin simple squamous epithelium that allows air entering the alveoli to exchange its gases with the blood passing through the capillaries.

Muscles of Respiration

Surrounding the lungs are sets of muscles that are able to cause air to be inhaled or exhaled from the lungs. The principal muscle of respiration in the human body is the diaphragm, a thin sheet of skeletal muscle that forms the floor of the thorax. When the diaphragm contracts, it moves inferiorly a few inches into the abdominal cavity, expanding the space within the thoracic cavity and pulling air into the lungs. Relaxation of the diaphragm allows air to flow back out the lungs during exhalation.

Between the ribs are many small intercostal muscles that assist the diaphragm with expanding and compressing the lungs. These muscles are divided into 2 groups: the internal intercostal muscles and the external intercostal muscles. The internal intercostal muscles are the deeper set of muscles and depress the ribs to compress the thoracic cavity and force air to be exhaled from the lungs. The external intercostals are found superficial to the internal intercostals and function to elevate the ribs, expanding the volume of the thoracic cavity and causing air to be inhaled into the lungs.

Physiology of the Respiratory System

Pulmonary Ventilation

Pulmonary ventilation is the process of moving air into and out of the lungs to facilitate gas exchange. The respiratory system uses both a negative pressure system and the contraction of muscles to achieve pulmonary ventilation. The negative pressure system of the respiratory system involves the establishment of a negative pressure gradient between the alveoli and the external atmosphere. The pleural membrane seals the lungs and maintains the lungs at a pressure slightly below that of the atmosphere when the lungs are at rest. This results in air following the pressure gradient and passively filling the lungs at rest. As the lungs fill with air, the pressure within the lungs rises until it matches the atmospheric pressure. At this point, more air can be inhaled by the contraction of the diaphragm and the external intercostal muscles, increasing the volume of the thorax and reducing the pressure of the lungs below that of the atmosphere again.

To exhale air, the diaphragm and external intercostal muscles relax while the internal intercostal muscles contract to reduce the volume of the thorax and increase the pressure within the thoracic cavity. The pressure gradient is now reversed, resulting in the exhalation of air until the pressures inside the lungs and outside of the body are equal. At this point, the elastic nature of the lungs causes them to recoil back to their resting volume, restoring the negative pressure gradient present during inhalation.

External Respiration

External respiration is the exchange of gases between the air filling the alveoli and the blood in the capillaries surrounding the walls of the alveoli. Air entering the lungs from the atmosphere has a higher partial pressure of oxygen and a lower partial pressure of carbon dioxide than does the blood in the capillaries. The difference in partial pressures causes the gases to diffuse passively along their pressure gradients from high to low pressure through the simple squamous epithelium lining of the alveoli. The net result of external respiration is the movement of oxygen from the air into the blood and the movement of carbon dioxide from the blood into the air. The oxygen can then be transported to the body's tissues while carbon dioxide is released into the atmosphere during exhalation.

Internal Respiration

Internal respiration is the exchange of gases between the blood in capillaries and the tissues of the body. Capillary blood has a higher partial pressure of oxygen and a lower partial pressure of carbon dioxide than the tissues through which it passes. The difference in partial pressures leads to the diffusion of gases along their pressure gradients from high to low pressure through the endothelium lining of the capillaries. The net result of internal respiration is the diffusion of oxygen into the tissues and the diffusion of carbon dioxide into the blood.

Transportation of Gases

The 2 major respiratory gases, oxygen and carbon dioxide, are transported through the body in the blood. Blood plasma has the ability to transport some dissolved oxygen and carbon dioxide, but most of the gases transported in the blood are bonded to transport molecules. Hemoglobin is an important transport molecule found in red blood cells that carries almost 99% of the oxygen in the blood. Hemoglobin can also carry a small amount of carbon dioxide from the tissues back to the lungs. However, the vast majority of carbon dioxide is carried in the plasma as bicarbonate ion. When the partial pressure of carbon dioxide is high in the tissues, the enzyme carbonic anhydrase catalyzes a reaction between carbon dioxide and water to form carbonic acid. Carbonic acid then dissociates into hydrogen ion and bicarbonate ion. When the partial pressure of carbon dioxide is low in the lungs, the reactions reverse and carbon dioxide is liberated into the lungs to be exhaled.

Homeostatic Control of Respiration

Under normal resting conditions, the body maintains a quiet breathing rate and depth called eupnea. Eupnea is maintained until the body's demand for oxygen and production of carbon dioxide rises due to greater exertion. Autonomic chemoreceptors in the body monitor the partial pressures of oxygen and carbon dioxide in the blood and send signals to the respiratory center of the brain stem. The respiratory center then adjusts the rate and depth of breathing to return the blood to its normal levels of gas partial pressures.

Health Issues Affecting the Respiratory System

When something impairs our ability to exchange carbon dioxide for oxygen, this is obviously a serious problem. Many health problems can cause respiratory problems, from allergies and asthma to pneumonia and lung cancer. The causes of these issues are just as varied—among them, infection (bacterial or viral), environmental exposure (pollution or cigarette smoke, for instance), genetic inheritance or a combination of factors. Sometimes the onset is so gradual, we don't seek medical attention until the condition has advanced. Sometimes, as with the genetic disorder called alpha-1 antitrypsin deficiency (A1AD), symptoms gradually set in and are often under-diagnosed or misdiagnosed. DNA health testing can screen you for genetic risk of A1AD.

Upper Respiratory System

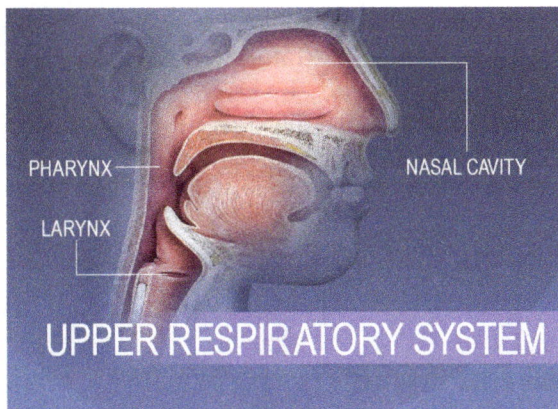

PHARYNX

NASAL CAVITY

LARYNX

UPPER RESPIRATORY SYSTEM

The upper respiratory system, or upper respiratory tract, consists of the nose and nasal cavity, the pharynx, and the larynx. These structures allow us to breathe and speak.

They warm and clean the air we inhale: mucous membranes lining upper respiratory structures trap some foreign particles, including smoke and other pollutants, before the air travels down to the lungs.

Nose and Nasal Cavities Provide Airways for Respiration

The nasal cavities are chambers of the internal nose. In front, the nostrils, or nares, create openings to the outside world. Air is inhaled through the nostrils and warmed as it moves further into the nasal cavities. Scroll-shaped bones, the nasal conchae, protrude and form spaces through which the air passes. The conchae swirl the air around to allow the air time to humidify, warm, and be cleaned before it enters the lungs. Epithelial cilia (commonly called "nose hair") and a mucous membrane line the inside of the cavities. The cilia, along with mucus produced by seromucous and other glands in the membrane, trap unwanted particles. Finally the filtered, warmed air passes out of the back of the nasal cavities into the nasopharynx, the uppermost part of the pharynx.

Paranasal Sinuses Surround the Nasal Cavities

The paranasal sinuses are four paired, air-filled cavities found inside bones of the skull. These sinuses are named for the skull bones that contain them: frontal, ethmoidal, sphenoidal, and maxillary. Mucosae line the paranasal sinuses and help to warm and humidify the air we inhale. When air enters the sinuses from the nasal cavities, mucus formed by the muscosae drains into the nasal cavities.

Pharynx Connects the Nasal and Oral Cavities to the Larynx and Esophagus

The pharynx, or throat, is shaped like a funnel. During respiration, it conducts air between the larynx and trachea (or "windpipe") and the nasal and the oral cavities. The pharynx includes three regions: The nasopharynx is posterior to the nasal cavity and serves only as a passageway for air. The oropharynx lies posterior to the oral cavity and contains the palatine tonsils. Both air and ingested food pass through the oropharynx and through the laryngopharynx below. The laryngopharynx lies posterior to the epiglottis and connects to the larynx (superiorly) and the esophagus (inferiorly). As we breathe, the epiglottis stays up and air passes freely between the laryngopharynx and the larynx.

Larynx and Vocal Cords Allow us to Breathe and Talk and Sing

The larynx connects the lower part of the pharynx, the laryngopharynx, to the trachea. It keeps the air passages open during breathing and digestion and is the key organ for producing sound. This larynx is comprised of nine cartilages. One, the epiglottis, is a lifesaver: Located on the posterior side of the larynx, the epiglottis closes like a trap door as we swallow. This action steers food down the esophagus and away from the windpipe. Inside the larynx are the vocal folds (or true vocal cords), which have elastic ligaments at their core. When we speak, yell, or sing, air coming up from the lungs and trachea vibrates the folds, producing the sound.

Hyoid is the only Bone in the Body that Doesn't Touch Another Bone

The U-shaped hyoid bone, located just under the chin, is an important contributor to both respiratory and digestive processes. The hyoid is attached to the tongue, and helps you to swallow at the start of digestion. In the respiratory system, structures that produce sound depend on the hyoid. The body and the greater horns of the bone serve as attachment points for neck muscles that raise and lower the larynx during speech (as well as during swallowing).

An upper respiratory tract infection, or URI, results when infectious agents enter the nose or mouth and travel into the upper part of the respiratory system, including the nose, trachea (breathing tube), and vocal cords. If this infection continues to travel, it can result in an infection of the lungs or a lower respiratory tract infection.

A cold is an infection of the nose and upper airways caused by a germ (virus). They are extremely common. An adult can expect 2-4 colds a year and a child can expect about 5-6 colds a year. Very young children in nursery school may get as many as 12 colds a year. Many different viruses can cause a cold. This is why colds come back (recur) and immunisation against colds is not possible.

Infections of the throat (larynx), or the main airway (trachea), or the airways going into the lungs (bronchi) are also common. These infections are sometimes called laryngitis, tracheitis, or bronchitis. Doctors often just use the term upper respiratory tract infection (URTI) to include any, or all, of these infections. Most URTIs are due to a viral infection.

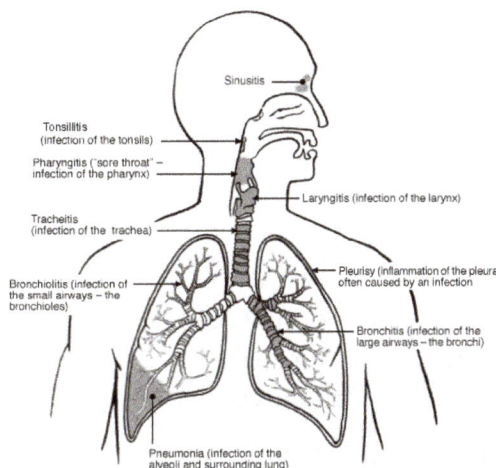

Figure: Sites of a range of respiratory infections

Symptoms of an Upper Respiratory Tract Infection

The common symptoms of a cold are a blocked (congested) nose, a runny nose and sneezing. At first there is a clear discharge (mucus) from the nose. This often becomes thick and yellow/green after 2-3 days. It may be difficult to sleep due to a blocked nose. You may feel generally unwell and tired and you may develop a cough or a mildly high temperature (a mild fever).

In other upper respiratory tract infections (URTIs), cough is usually the main symptom. Other symptoms include fever, headache, aches and pains.

Symptoms are typically at their worst after 2-3 days and then gradually clear. However, the cough may carry on after the infection has gone. This is because swelling (inflammation) in the airways, caused by the infection, can take a while to settle. It may take 2-3 weeks, after other symptoms have gone, for a cough to clear completely.

Treatments for an Upper Respiratory Tract Infection

A main aim of treatment for an upper respiratory tract infection (URTI) is to ease symptoms whilst your immune system clears the infection. One or more of the following may be helpful:

- Taking paracetamol or ibuprofen to reduce a high temperature (fever) and to ease any aches, pains and headaches. Follow the instructions given with the medicine carefully and do not take more than the advised dose. (Only give these medicines to children under the age of 5 years if they have a fever or appear distressed.)

- Having plenty to drink if you have a fever, to prevent mild lack of fluid in the body (dehydration). As long as you do not have a fever, there is no evidence that drinking more fluid than usual makes a difference.

- If you smoke, you should try to stop for good. URTIs and serious lung diseases tend to last longer in smokers.

- Steam inhalation: There is not very much evidence that this helps; however, some people find it useful. It is very important to be careful to avoid burns and scalds, particularly with children. A safe way of inhaling steam is to sit in the bathroom with the door closed, while running a hot shower to make the room steamy.

- Vapour rubs: Vapour rubs can be bought in pharmacies and supermarkets. Some people find they help with a stuffy nose. Rub the vapour on to the chest and/or back of the person with the cold, but avoid the area under the nose.

- Sucking sore throat lozenges (available from pharmacies and supermarkets) or boiled sweets may help ease a sore throat.

- Warm drinks with honey and lemon may help to ease a sore throat. (Do not give honey to babies less than 1 year old as it is not known if this is safe.)

- Salt (saline) nose drops: These are nose drops made of a salty solution, which may help clear a blocked nose. They are sometimes helpful for babies who are having difficulty breathing through a blocked nose as they feed.

Prevention for Upper Respiratory Tract Infections

Prevention is difficult. Many germs (viruses) can cause an upper respiratory tract infection (URTI). Also, many viruses that cause URTIs are in the air, which you cannot avoid. However, the following are suggestions that may reduce the risk of catching a URTI or of passing one on, if you have one:

- If you have a URTI do not get too close to others - for example, kissing, hugging, etc.

- If you have a URTI, wash your hands often with soap and water. Many viruses are passed on by touch, especially from hands that are contaminated with a virus.

- Avoid sharing towels, flannels, etc, if you have a URTI, or with anyone who has a URTI.

- For children, discourage the sharing of toys belonging to a child with a URTI. If your child has a URTI, consider washing toys with soapy water after use.

Lower Respiratory System

The lower respiratory system consists of the trachea, bronchi and lungs. The air passages are lined with mucous membrane composed mainly of ciliated epithelium. Cilia constantly clean the tract and carry foreign matter upwards for swallowing or expectoration.

Trachea

The trachea (windpipe) extends from the laryngopharynx at the level of the cricoid cartilage at the top to the carina (also called the tracheal bifurcation). C-shaped cartilage rings reinforce and protect the trachea to prevent it from collapsing. The carina is a ridge-shaped structure at the level of T6 or T7. The carina possesses sensory nerve endings which cause coughing if food or water is inhaled accidently.

Bronchi

The primary bronchi begin at the carina. The right primary bronchus - shorter, wider and more vertical than the left - supplies air to the right lung. The left primary bronchus delivers air to the left lung. Along with blood vessels, nerves, and lymphatics, the primary bronchi enter the lungs at the hilum. Located behind the heart, the hilum is a slit on the lung's medial surface.

Secondary Bronchi

Each primary bronchus divides to form secondary bronchi. In each lung, one secondary bronchus goes into each lobe which means that the right lung has three secondary bronchi and the left lung has two.

Branching Out

Each lobar bronchus enters a lobe in each lung. Within its lobe, each of the lobar bronchi branches into segmental bronchi (tertiary bronchi). The segments continue to branch into smaller and smaller bronchi, finally branching into bronchioles. The larger bronchi consist of cartilage, smooth muscle and epithelium. As the bronchi become smaller, they lose cartilage and then smooth muscle. Ultimately, the smallest bronchioles consist of just a single layer of epithelial cells.

Respiratory Bronchioles

Each bronchiole includes terminal bronchioles and the alveolar sac - the chief respiratory unit for gas exchange. Within the acinus, terminal bronchioles branch into yet smaller respiratory bronchioles. The respiratory bronchioles feed directly into alveoli at sites along their walls.

Alveoli

The respiratory bronchioles eventually become alveolar ducts, which terminate in clusters of alveoli surrounded by capillaries (alveolar sacs). Gas exchange takes place through the alveoli.

Alveolar walls contain two basic epithelial cell types:

- Type I cells are the most abundant. It is across these thin, flat, squamous cells that gas exchange occurs.

- Type II cells secrete surfactant, a substance that coats the alveolus and reduces surface tension. This allows the alveoli to remain inflated so that gas exchange can occur by diffusion. Surfactant is formed relatively late in foetal life; thus premature infants born without adequate amounts experience respiratory distress and may die.

Lungs

The lungs are a pair of breathing organs located with the chest which remove carbon dioxide from and bring oxygen to the blood. There is a right and left lung.

Structure

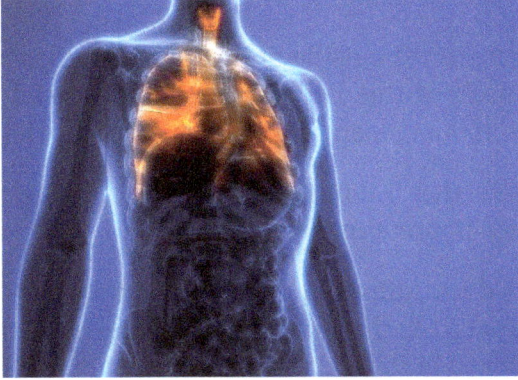

The lungs not only enable us to breathe and talk, but they also support the cardiovascular system and help maintain pH in the body, among others.

The lungs are located in the chest, behind the rib cage on either side of the heart. They are roughly conical in shape with a rounded point at their apex and a flatter base where they meet the diaphragm.

Although they are a pair, the lungs are not equal in size and shape.

The left lung has an indentation bordering where the heart resides, called the cardiac notch. The right lung is shorter to allow space for the liver below.

Overall, the left lung has a slightly smaller weight and capacity than the right.

The lungs are surrounded by two membranes, known as the pulmonary pleurae. The inner layer directly lines the outer surface of the lungs, and the outer layer is attached to the inner wall of the rib cage.

The space between the two membranes is filled with pleural fluid.

Function

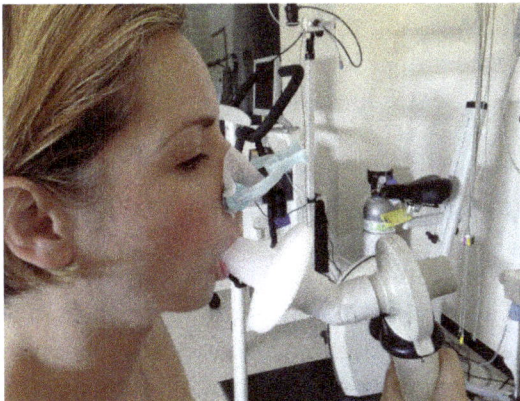

A spirometry test can show how well the lungs are working.

The lungs' main role is to bring in air from the atmosphere and pass oxygen into the bloodstream. From there, it circulates to the rest of the body.

Help is required from structures outside of the lungs in order to breathe properly. To breathe, we use the muscle of the diaphragm, the intercostal muscles (between the ribs), the muscles of the abdomen, and sometimes even muscles in the neck.

The diaphragm is a muscle that is domed at the top and sits below the lungs. It powers most of the work involved in breathing.

As it contracts, it moves down, allowing more space in the chest cavity and increasing the lungs' capacity to expand. As the chest cavity volume increases, the pressure inside goes down, and air is sucked in through the nose or mouth and down into the lungs.

As the diaphragm relaxes and returns to its resting position, the lung volume decreases because the pressure inside the chest cavity goes up, and the lungs expel the air.

The lungs are like bellows. As they expand, air is sucked in for oxygen. As they compress, the exchanged carbon dioxide waste is pushed back out during exhalation.

When air enters the nose or mouth, it travels down the trachea, also called the windpipe. After this, it reaches a section called the carina. At the carina, the windpipe splits into two, creating two mainstem bronchi. One leads to the left lung and the other to the right lung.

Lower Respiratory Tract Infection

The infection which occurs in the lower respiratory tract of human body is usually termed by the doctors as lower respiratory tract infection. The infections begin from the lower larynx and can also attack the bronchi and even the whole lungs. The common illnesses are bronchiolitis, pneumonia, bronchitis and flu.

The viruses which cause lower respiratory tract infection spread over the tract slowly. The diseases are mostly contagious in nature. Through direct contact with the patient of by inhaling the air which contains the germs you can get the same illness. Like the upper respiratory tract infections, lower respiratory tract infection spreads very fast and these infections cause greater harm to the human body than the upper respiratory infections.

Causes of Lower Respiratory Tract Infection

The primary cause of lower respiratory tract infection are the various kinds of viruses that attack our system. The viruses that enter our body often take the shape of structural proteins and hence can pass through unrecognised by the immunity system. Also the viruses secrete some of the toxins which are stronger than our immune system and thus our immunity drops down further and we get attacked by them. Some of these viruses enter in their 1st stage and thus they have a cover which the immunity system cannot penetrate or even identify as danger. Then the viruses incubate for a certain period of days and then the fully grown viruses attack the system. Within the incubation period the viruses learn about the immune system of the body and thus can fight them easily to spread the disease.

Common lower respiratory tract infections include:

- Bronchitis is a common lower respiratory tract infection. There are 2 kinds of Bronchitis; acute and chronic. The virus swells the bronchial tubes which causes difficulty in breathing, thus the infection affects the airways. The main virus which causes the disease is the same that causes the flu. Often the viruses of influenza and pneumonia attack the patient at the same time. About 4% of the people in a population of 1000 are affected by this virus. Whether the disease is chronic or acute is decided by the stage of the virus and the presentation of its structure.

- Pneumonia is another lower respiratory tract infection. It is caused by the virus streptococcus pneumoniae. The virus causes a great damage to the lung and the mortality rate is 25% of the patients affected by the virus. If a child below the age of 5 is affected, then he or she might not survive at all. Though this is generally not contagious but if your immunity is low then being in direct contact with the cough discharges may pass on the virus on you.

- Flu is caused by the influenza viruses and affects both the upper and the lower respiratory tracts.

- Bacterial meningitis caused by the virus Neisseria meningitides can cause lower respiratory tract infection.

- Scarlet fever cause by Group A strep.

- Tuberculosis cause by the bacteria mycobacterium tuberculosis which causes a consistent damage to the lungs.

- Bronchiolitis is also a lower respiratory tract infection. It is caused by RSV or respiratory syncytial virus. It mainly affects the respiratory tracts and the airways of little children.

Signs and Symptoms of Lower Respiratory Tract Infection

The primary signs and symptoms of any lower respiratory tract infection is coughing. Often the coughing is followed by mucus discharges. As you cough you, the phlegm may come out of your mouth. This is the sign of deterioration and thus cannot be ignored as common cold. Soon the other signs and symptoms like high temperature, dizziness, feeling a pressure or heaviness in your chest, shortness of breathing or breathing at an increased rate follow. Also blocking of nose, watering of nose and increased heart beat can be the symptoms of lower respiratory tract infection. Asthma attacks may also occur. As the lung gets affected, the breathing problem becomes acute. In case of pneumonia, the body temperature rises too high that it may also affect the brain. Your immune system becomes very weak, you start feeling weak hence. The viruses may also attack your nervous system which can cause sclerosis.

The health of the patient worsens slowly over a period of a week. In the beginning the symptoms may resemble that of a common cold or flu but then body aches begin and you feel like you are not being able to breathe and your fever goes on increasing.

References

- Upper-respiratory-system, respiratory: visiblebody.com, Retrieved 29 April 2018

- Upper-respiratory-tract-infection-uri-causes-symptoms-treatment: study.com, Retrieved 19 May 2018

- Common-cold-upper-respiratory-tract-infections, cough-leaflet: patient.info, Retrieved 13 July 2018

- The-lower-respiratory-tract-made-incredibly-easy-5042166: nursingtimes.net, Retrieved 17 April 2018

- Causes-and-symptoms-of-lower-respiratory-tract-infection, chest-pain: epainassist.com, Retrieved 29 June 2018

8

Urinary System

The urinary system of the human body consists of a pair of kidneys, ureters, the urinary bladder and the urethra. All these organs of the human urinary system have been extensively covered in this chapter.

The urinary system consists of the kidneys, ureters, urinary bladder, and urethra. The kidneys filter the blood to remove wastes and produce urine. The ureters, urinary bladder, and urethra together form the urinary tract, which acts as a plumbing system to drain urine from the kidneys, store it, and then release it during urination. Besides filtering and eliminating wastes from the body, the urinary system also maintains the homeostasis of water, ions, pH, blood pressure, calcium and red blood cells.

Urinary System Anatomy

Kidneys

The kidneys are a pair of bean-shaped organs found along the posterior wall of the abdominal cavity. The left kidney is located slightly higher than the right kidney because the right side of the liver is much larger than the left side. The kidneys, unlike the other organs of the abdominal cavity, are located posterior to the peritoneum and touch the muscles of the back. The kidneys are surrounded by a layer of adipose that holds them in place and protects them from physical damage. The kidneys filter metabolic wastes, excess ions, and chemicals from the blood to form urine.

Ureters

The ureters are a pair of tubes that carry urine from the kidneys to the urinary bladder. The ureters are about 10 to 12 inches long and run on the left and right sides of the body parallel to the vertebral column. Gravity and peristalsis of smooth muscle tissue in the walls of the ureters move urine toward the urinary bladder. The ends of the ureters extend slightly into the urinary bladder and are sealed at the point of entry to the bladder by the ureterovesical valves. These valves prevent urine from flowing back towards the kidneys.

Urinary Bladder

The urinary bladder is a sac-like hollow organ used for the storage of urine. The urinary bladder is located along the body's midline at the inferior end of the pelvis. Urine entering the urinary bladder from the ureters slowly fills the hollow space of the bladder and stretches its elastic walls. The walls of the bladder allow it to stretch to hold anywhere from 600 to 800 milliliters of urine.

Urethra

The urethra is the tube through which urine passes from the bladder to the exterior of the body. The female urethra is around 2 inches long and ends inferior to the clitoris and superior to the vaginal opening. In males, the urethra is around 8 to 10 inches long and ends at the tip of the penis. The urethra is also an organ of the male reproductive system as it carries sperm out of the body through the penis.

The flow of urine through the urethra is controlled by the internal and external urethral sphincter muscles. The internal urethral sphincter is made of smooth muscle and opens involuntarily when the bladder reaches a certain set level of distention. The opening of the internal sphincter results in the sensation of needing to urinate. The external urethral sphincter is made of skeletal muscle and may be opened to allow urine to pass through the urethra or may be held closed to delay urination.

Urinary System Physiology

Maintenance of Homeostasis

The kidneys maintain the homeostasis of several important internal conditions by controlling the excretion of substances out of the body.

Ions

The kidney can control the excretion of potassium, sodium, calcium, magnesium, phosphate, and chloride ions into urine. In cases where these ions reach a higher than normal concentration, the kidneys can increase their excretion out of the body to return them to a normal level. Conversely, the kidneys can conserve these ions when they are present in lower than normal levels by allowing the ions to be reabsorbed into the blood during filtration.

pH

The kidneys monitor and regulate the levels of hydrogen ions (H^+) and bicarbonate ions in the blood to control blood pH. H^+ ions are produced as a natural byproduct of the metabolism of dietary proteins and accumulate in the blood over time. The kidneys excrete excess H^+ ions into urine for elimination from the body. The kidneys also conserve bicarbonate ions, which act as important pH buffers in the blood.

Osmolarity

The cells of the body need to grow in an isotonic environment in order to maintain their fluid and electrolyte balance. The kidneys maintain the body's osmotic balance by controlling the amount of water that is filtered out of the blood and excreted into urine. When a person consumes a large amount of water, the kidneys reduce their reabsorption of water to allow the excess water to be excreted in urine. This results in the production of dilute, watery urine. In the case of the body being dehydrated, the kidneys reabsorb as much water as possible back into the blood to produce highly concentrated urine full of excreted ions and wastes. The changes in excretion of water are controlled by antidiuretic hormone (ADH). ADH is produced in the hypothalamus and released by the posterior pituitary gland to help the body retain water.

Blood Pressure

The kidneys monitor the body's blood pressure to help maintain homeostasis. When blood pressure is elevated, the kidneys can help to reduce blood pressure by reducing the volume of blood in the body. The kidneys are able to reduce blood volume by reducing the reabsorption of water into the blood and producing watery, dilute urine. When blood pressure becomes too low, the kidneys can produce the enzyme renin to constrict blood vessels and produce concentrated urine, which allows more water to remain in the blood.

Filtration

Inside each kidney are around a million tiny structures called nephrons. The nephron is the functional unit of the kidney that filters blood to produce urine. Arterioles in the kidneys deliver blood to a bundle of capillaries surrounded by a capsule called a glomerulus. As blood flows through the glomerulus, much of the blood's plasma is pushed out of the capillaries and into the capsule, leaving the blood cells and a small amount of plasma to continue flowing through the capillaries. The liquid filtrate in the capsule flows through a series of tubules lined with filtering cells and surrounded by capillaries. The cells surrounding the tubules selectively absorb water and substances from the filtrate in the tubule and return it to the blood in the capillaries. At the same time, waste products present in the blood are secreted into the filtrate. By the end of this process, the filtrate in the tubule has become urine containing only water, waste products, and excess ions. The blood exiting the capillaries has reabsorbed all of the nutrients along with most of the water and ions that the body needs to function.

Storage and Excretion of Wastes

After urine has been produced by the kidneys, it is transported through the ureters to the urinary bladder. The urinary bladder fills with urine and stores it until the body is ready for its excretion. When the volume of the urinary bladder reaches anywhere from 150 to 400 milliliters, its walls begin to stretch and stretch receptors in its walls send signals to the brain and spinal cord. These signals result in the relaxation of the involuntary internal urethral sphincter and the sensation of needing to urinate. Urination may be delayed as long as the bladder does not exceed its maximum volume, but increasing nerve signals lead to greater discomfort and desire to urinate.

Urination is the process of releasing urine from the urinary bladder through the urethra and out of the body. The process of urination begins when the muscles of the urethral sphincters relax, allowing urine to pass through the urethra. At the same time that the sphincters relax, the smooth muscle in the walls of the urinary bladder contract to expel urine from the bladder.

Production of Hormones

The kidneys produce and interact with several hormones that are involved in the control of systems outside of the urinary system.

Calcitriol

Calcitriol is the active form of vitamin D in the human body. It is produced by the kidneys from precursor molecules produced by UV radiation striking the skin. Calcitriol works together with parathyroid hormone (PTH) to raise the level of calcium ions in the bloodstream. When the level of calcium ions in the blood drops below a threshold level, the parathyroid glands release PTH, which in turn stimulates the kidneys to release calcitriol. Calcitriol promotes the small intestine to absorb calcium from food and deposit it into the bloodstream. It also stimulates the osteoclasts of the skeletal system to break down bone matrix to release calcium ions into the blood.

Erythropoietin

Erythropoietin, also known as EPO, is a hormone that is produced by the kidneys to stimulate the production of red blood cells. The kidneys monitor the condition of the blood that passes through their capillaries, including the oxygen-carrying capacity of the blood. When the blood becomes hypoxic, meaning that it is carrying deficient levels of oxygen, cells lining the capillaries begin producing EPO and release it into the blood-stream. EPO travels through the blood to the red bone marrow, where it stimulates hematopoietic cells to increase their rate of red blood cell production. Red blood cells contain hemoglobin, which greatly increases the blood's oxygen-carrying capacity and effectively ends the hypoxic conditions.

Renin

Renin is not a hormone itself, but an enzyme that the kidneys produce to start the re-nin-angiotensin system (RAS). The RAS increases blood volume and blood pressure in response to low blood pressure, blood loss, or dehydration. Renin is released into the blood where it catalyzes angiotensinogen from the liver into angiotensin I. Angiotensin I is further catalyzed by another enzyme into Angiotensin II.

Angiotensin II stimulates several processes, including stimulating the adrenal cor-tex to produce the hormone aldosterone. Aldosterone then changes the function of the kidneys to increase the reabsorption of water and sodium ions into the blood, increasing blood volume and raising blood pressure. Negative feedback from in-creased blood pressure finally turns off the RAS to maintain healthy blood pressure levels.

Kidney

The kidneys are a pair of bean-shaped organs on either side of your spine, below your ribs and behind your belly. Each kidney is about 4 or 5 inches long, roughly the size of a large fist.

The kidneys' job is to filter your blood. They remove wastes, control the body's fluid balance, and keep the right levels of electrolytes. All of the blood in your body passes through them several times a day.

Blood comes into the kidney, waste gets removed, and salt, water, and minerals are adjusted, if needed. The filtered blood goes back into the body. Waste gets turned into urine, which collects in the kidney's pelvis - a funnel-shaped structure that drains down a tube called the ureter to the bladder.

Each kidney has around a million tiny filters called nephrons. You could have only 10% of your kidneys working, and you may not notice any symptoms or problems.

If blood stops flowing into a kidney, part or all of it could die. That can lead to kidney failure.

Nephrons

Nephrons are the most important part of each kidney. They take in blood, metabolize nutrients, and help pass out waste products from filtered blood. Each kidney has about 1 million nephrons. Each has its own internal set of structures.

Renal Corpuscle

After blood enters a nephron, it goes into the renal corpuscle, also called a Malpighian body. The renal corpuscle contains two additional structures:

- Glomerulus: This is a cluster of capillaries that absorb protein from blood traveling through the renal corpuscle.

- Bowman capsule: The remaining fluid, called capsular urine, passes through the Bowman capsule into the renal tubules.

Renal Tubules

The renal tubules are a series of tubes that begin after the Bowman capsule and end at collecting ducts.

Each tubule has several parts:

- Proximal convoluted tubule: This section absorbs water, sodium, and glucose back into the blood.

- Loop of Henle: This section further absorbs potassium, chloride, and sodium into the blood.

- Distal convoluted tubule: This section absorbs more sodium into the blood and takes in potassium and acid.

By the time fluid reaches the end of the tubule, it's diluted and filled with urea. Urea is byproduct of protein metabolism that's released in urine.

Renal Cortex

The renal cortex is the outer part of the kidney. It contains the glomerulus and convoluted tubules.

The renal cortex is surrounded on its outer edges by the renal capsule, a layer of fatty tissue. Together, the renal cortex and capsule house and protect the inner structures of the kidney.

Renal Medulla

The renal medulla is the smooth, inner tissue of the kidney. It contains the loop of Henle as well as renal pyramids.

Renal Pyramids

Renal pyramids are small structures that contain strings of nephrons and tubules. These tubules transport fluid into the kidney. This fluid then moves away from the nephrons toward the inner structures that collect and transport urine out of the kidney.

Collecting Ducts

There's a collecting duct at the end of each nephron in the renal medulla. This is where filtered fluids exit the nephrons.

Once in the collecting duct, the fluid moves on to its final stops in the renal pelvis.

Renal Pelvis

The renal pelvis is a funnel-shaped space in the innermost part of the kidney. It functions as a pathway for fluid on its way to the bladder.

Calyces

The first part of the renal pelvis contains the calyces. These are small cup-shaped spaces that collect fluid before it moves into the bladder. This is also where extra fluid and waste become urine.

Hilum

The hilum is a small opening located on the inner edge of the kidney, where it curves inward to create its distinct beanlike shape. The renal pelvis passes through it, as well as the:

- Renal artery: This brings oxygenated blood from the heart to the kidney for filtration.
- Renal vein: This carries filtered blood from the kidneys back to the heart.

Functions of the Kidney

The kidneys are important for excretion, however they play more vital roles to life than we realize. The kidneys perform the following functions:

1. Regulation of Arterial Blood Pressure

The kidneys excrete a great amount of sodium and water. They secrete an enzyme called rennin that activates the renin-angiotensin system that control blood pressure and sodium concentration.

2. Regulation of Water and Electrolyte Balance

In order for our bodies to maintain a balance, the amount of water and electrolyte we take in must be the same amount we excrete. If we take in more than we excrete, the kidneys will work to regulate the excess and bring the body to a balance.

3. Excretion of Metabolic Waste Products and Foreign Chemicals

Our bodies, each day, form waste products that it does not need. These products include urea, creatinine, uric acid and waste products from the breakdown of hemoglobin that give the urine the color it has.

If these products are not eliminated from the body as fast as they are produced they become very harmful to the body. The kidneys remove these waste products so, our body can function normally.

4. Regulation of Red Blood Cell Production

Our kidneys secrete erythropoietin, a hormonea gland, or an organ in one part of the body that affects cells in other parts of the organism. Generally, only a small amount of hormone is required to alter cell metabolism. In essence, it is a chemical messenger that t" which stimulates the production of red blood cells in the body. If the kidney is removed or severely damaged, then we may not be able to produce RBC and severe anemia will develop as a result of the decrease in the production of erythropoietin.

5. Regulation of Vitamin D Production

Our kidneys produce the active form of vitamin D, calcitriol. We get vitamin D from sunlight or from ingested vitamin. These types of vitamin D are in their inactive form. The kidneys are needed to convert them into their active forms.

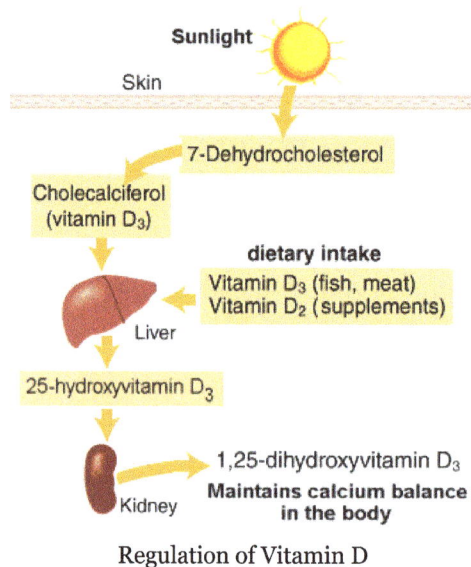

Regulation of Vitamin D

The active form (calcitriol) is necessary for normal calcium absorption in the bone. Therefore, if the kidneys are damaged there will be a decreased level in calcitriol which then will lead to a reduction in calcium absorption, and thus low calcium in the body leading to bone disorders.

6. Gluconeogenesis

If we should stop eating carbohydrates (our main source of glucose) for a day, our bodies would begin to form new glucose from the amino acid in the proteins we intake. This process is known as gluconeogenesis.

So, when the kidneys are damaged, this function is crippled, and this can cause death within a few days.

7. Regulation of Acid Base Balance

The kidneys are able to perform these functions through the work of the nephrons and the collecting tubules.

Kidney Conditions

- Pyelonephritis (infection of kidney pelvis): Bacteria may infect the kidney, usually causing back pain and fever. A spread of bacteria from an untreated bladder infection is the most common cause of pyelonephritis.

- Glomerulonephritis: An overactive immune system may attack the kidney, causing inflammation and some damage. Blood and protein in the urine are common problems that occur with glomerulonephritis. It can also result in kidney failure.

- Kidney stones (Nephrolithiasis): Minerals in urine form crystals (stones), which may grow large enough to block urine flow. It's considered one of the most painful conditions. Most kidney stones pass on their own, but some are too large and need to be treated.

- Nephrotic syndrome: Damage to the kidneys causes them to spill large amounts of protein into the urine. Leg swelling (edema) may be a symptom.

- Polycystic kidney disease: A genetic condition resulting in large cysts in both kidneys that hinder their work.

- Acute renal failure (kidney failure): A sudden worsening in how well your kidneys work. Dehydration, a blockage in the urinary tract, or kidney damage can cause acute renal failure, which may be reversible.

- Chronic renal failure: A permanent partial loss of how well your kidneys work. Diabetes and high blood pressure are the most common causes.

- End-stage renal disease (ESRD): Complete loss of kidney strength, usually due to progressive chronic kidney disease. People with ESRD require regular dialysis for survival.

- Papillary necrosis: Severe damage to the kidneys can cause chunks of kidney tissue to break off internally and clog the kidneys. If untreated, the resulting damage can lead to total kidney failure.

- Diabetic nephropathy: High blood sugar from diabetes progressively damages the kidneys, eventually causing chronic kidney disease. Protein in the urine (nephrotic syndrome) may also result.

- Hypertensive nephropathy: Kidney damage caused by high blood pressure. Chronic renal failure may eventually result.

- Kidney cancer: Renal cell carcinoma is the most common cancer affecting the kidney. Smoking is the most common cause of kidney cancer.

- Interstitial nephritis: Inflammation of the connective tissue inside the kidney, often causing acute renal failure. Allergic reactions and drug side effects are the usual causes.

- Minimal change disease: A form of nephrotic syndrome in which kidney cells look almost normal under the microscope. The disease can cause significant leg swelling (edema). Steroids are used to treat minimal change disease.

- Nephrogenic diabetes insipidus: The kidneys lose the ability to concentrate the urine, usually due to a drug reaction. Although it's rarely dangerous, diabetes insipidus causes constant thirst and frequent urination.

- Renal cyst: A hollowed-out space in the kidney. Isolated kidney cysts often happen as people age, and they almost never cause a problem. Complex cysts and masses can be cancerous.

Ureter

The ureters are a pair of small tubes that connect the kidneys to the urinary bladder. They form a vital link in the urinary tract by allowing urine to drain from the kidneys to be stored in the bladder.

Anatomy

The ureters are thin, muscular tubes found within the abdominopelvic cavity. Each is about 10 to 12 inches (25-30 cm) long, with the left ureter being slightly longer than the right due to the left kidney lying superior to the right kidney. The width of the ureters is only a few millimeters in diameter and varies from one end to the other — wider near the kidneys and narrower near the urinary bladder.

Each ureter begins at the kidney where the renal pelvis exits through the renal hilus. The ureter begins as a tube about 10 mm in diameter as it exits the kidney, but it tapers to only 1-2 mm as it descends through the abdominopelvic cavity. Like the kidneys, the ureters are retroperitoneal organs, which pass through the walls of the abdominopelvic cavity posterior to the peritoneum. Initially the ureters point medially, but quickly turn inferiorly and descend toward the pelvis. Within the pelvis, the ureters turn anteriorly to pass over the common iliac arteries and veins before turning posteriorly to resume their original courses. At the base of the pelvis, the ureters once again turn anteriorly to approach the urinary bladder and enter the bladder on its posterior wall. The ureters pass obliquely through the wall of the urinary bladder and end at the ureteral openings about 1 inch (2 cm) within the bladder.

While the ureters may appear to be simple tubes for urine to drain into the urinary bladder, they actually play active roles in the urinary system. Under most circumstances, gravity is able to pull urine down from the kidneys to the urinary bladder. However, gravity would not be able to move urine to the bladder when one is lying down or in freefall (such as an astronaut in space). To ensure the unidirectional flow of urine, the ureters use waves of smooth muscle contraction, known as peristalsis, to conduct urine to the urinary bladder. Once urine has entered the urinary bladder, the pressure of the inflated bladder squeezes the ends of the ureters, preventing urine from flowing back towards the kidneys. These processes for keeping urine flowing away from the kidneys help to eliminate waste in the most efficient way possible and help to prevent bacterial infections from reaching the kidneys.

Urinary Bladder

The urinary bladder is a muscular sac in the pelvis, just above and behind the pubic bone. When empty, the bladder is about the size and shape of a pear.

Urine is made in the kidneys and travels down two tubes called ureters to the bladder. The bladder stores urine, allowing urination to be infrequent and controlled. The bladder is lined by layers of muscle tissue that stretch to hold urine. The normal capacity of the bladder is 400-600 mL.

During urination, the bladder muscles squeeze, and two sphincters (valves) open to allow urine to flow out. Urine exits the bladder into the urethra, which carries urine out of the body. Because it passes through the penis, the urethra is longer in men (8 inches) than in women (1.5 inches).

The urinary bladder is made of several distinct tissue layers:

- The innermost layer of the bladder is the mucosa layer that lines the hollow lumen. Unlike the mucosa of other hollow organs, the urinary bladder is lined with transitional epithelial tissue that is able to stretch significantly to accommodate large volumes of urine. The transitional epithelium also provides protection to the underlying tissues from acidic or alkaline urine.

- Surrounding the mucosal layer is the submucosa, a layer of connective tissue with blood vessels and nervous tissue that supports and controls the surrounding tissue layers.

- The visceral muscles of the muscularis layer surround the submucosa and provide the urinary bladder with its ability to expand and contract. The muscularis is commonly referred to as the detrusor muscle and contracts during urination to expel urine from the body. The muscularis also forms the internal urethral sphincter, a ring of muscle that surrounds the urethral opening and holds urine in the urinary bladder. During urination, the sphincter relaxes to allow urine to flow into the urethra.

The urinary bladder sits in a unique position inferior to the peritoneum, a membrane that lines most of the abdominopelvic cavity. Due to its position, the outermost layer of the superior urinary bladder is made of serous membrane continuous with the peritoneum. Serous membrane provides protection to the bladder from friction between organs in the abdominopelvic cavity. The surface of the lateral and inferior sides of the urinary bladder forms a layer of loose connective tissue known as the adventitia. The adventitia loosely connects the urinary bladder to the surrounding tissues of the pelvis.

The urinary bladder functions as a storage vessel for urine to delay the frequency of urination. It is one of the most elastic organs of the body and is able to increase its volume greatly to accommodate between 600 to 800 ml of urine at maximum capacity. Transitional epithelium, elastic fibers, and visceral muscle tissue in the walls of the urinary bladder contribute to its distensibility and elasticity, allowing it to easily stretch and return to its original size several times each day.

It also helps to expel urine from the body during urination by contraction of the detrusor muscle and the relaxation of the internal urethral sphincter. Another separate muscle, the external urethral sphincter, surrounds the urethra just inferior to the bladder and helps to control and delay urination through its contraction. The external urethral sphincter is a skeletal muscle and therefore allows for the voluntary control of the urination reflex.

Functions of the Bladder

The bladder largely serves two functions:

- Temporary store of urine – The bladder is a hollow organ. The walls are very distensible, with a folded internal lining (known as rugae), this allows it to hold up to 600ml.

- Assists in the expulsion of urine – During voiding, the musculature of the bladder contracts, and the sphincters relax.

Shape of the Bladder

The morphological appearance of the bladder varies with filling. When full, it exhibits an oval shape, and when empty it is flattened by the overlying intestines.

The important external features are the apex, body, fundus and neck:

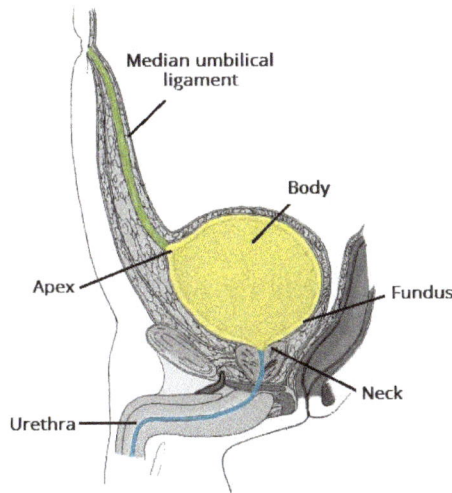

Figure: Sagittal section of the male pelvis. The external anatomical features of the bladder

Apex – This is located superiorly, pointing towards the pubic symphysis. It is connected to the umbilicus by the median umbilical ligament (a remnant of the urachus).

Body – The main part of the bladder, located between the apex and the fundus

Fundus (or base) – Located posteriorly. It is triangular-shaped, with the tip of the triangle pointing backwards.

Neck – Formed by the convergence of the fundus and the two inferolateral surfaces. This structure joins the bladder to the urethra.

Urine enters the bladder by the left and right ureters, and exits via the urethra. Internally, these orifices are marked by the trigone - a triangular area located within the fundus. In contrast to the rest of the internal bladder, the trigone has smooth walls.

There are two sphincters controlling the outflow of urine; the internal and external urethral sphincters.

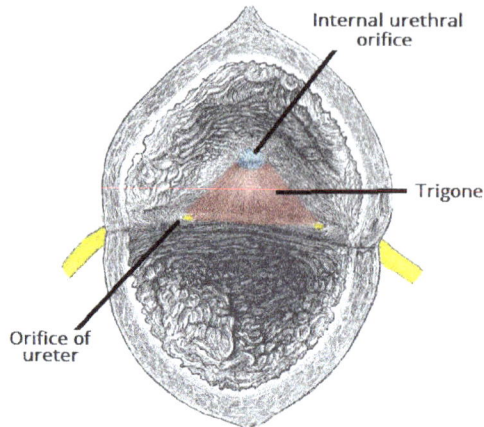

Figure: The internal surface of the bladder, highlighting the trigone.

Musculature

The bladder musculature, and the coordination of its action, plays a key role in the functions of the bladder.

In order to contract during micturition, the bladder wall contains specialised smooth muscle, known as detrusor muscle. Its fibres are orientated in three directions, thus retaining structural integrity when stretched. It receives innervation from both the sympathetic and parasympathetic nervous systems.

There are also two muscular sphincters located in the urethra:

- Internal urethral sphincter:

  o Male – consists of circular smooth fibres, which are under autonomic control. It is thought to prevent seminal regurgitation during ejaculation.

  o Females – thought to be a functional sphincter (i.e. no sphincteric muscle present). It is formed by the anatomy of the bladder neck and proximal urethra.

- External urethral sphincter – has the same structure in both sexes. It is skeletal muscle, and under voluntary control. During micturition, it relaxes to allow urine flow.

Vasculature

The bladder primarily receives its vasculature from the internal iliac vessels.

Arterial supply is delivered by the superior vesical branch of the internal iliac artery. In males, this is supplemented by the inferior vesical artery, and in females by the vaginal arteries. In both sexes, the obturator and inferior gluteal arteries also contribute small branches.

Venous drainage is achieved by the vesical venous plexus, which empty into the internal iliac vein (also known as the hypogastric vein).

Nervous Supply

Neurological control is complex, with the bladder receiving input from both the autonomic (sympathetic and parasympathetic) and somatic arms of the nervous system:

- The sympathetic nervous system communicates with the bladder via the hypogastric nerve (T12 – L2). It causes relaxation of the detrusor muscle. These functions promote urine retention.

- The parasympathetic nervous system communicates with the bladder via the pelvic nerve (S2-S4). Increased signals from this nerve causes contraction of the detrusor muscle. This stimulates micturition.

- The somatic nervous supply gives us voluntary control over micturition. It innervates the external urethral sphincter, via the pudendal nerve (S2-S4). It can cause it to constrict (storage phase) or relax (micturition).

In addition to the efferent nerves supplying the bladder, there are sensory (afferent) nervesthat report to the brain. They are found in the bladder wall and signal the need to urinate when the bladder becomes full.

Bladder Stretch Reflex

The bladder stretch reflex is a primitive spinal reflex, in which micturition is stimulated in response to stretch. It is analogous to a muscle spinal reflex, such as the patella reflex.

During toilet training in infants, this spinal reflex is overridden by the higher centers of the brain, to give voluntary control over micturition.

The reflex arc:

- Bladder fills with urine, and the bladder walls stretch. Sensory nerves detect stretch and transmit this information to the spinal cord.

- Interneurons within the spinal cord relay the signal to the parasympathetic efferents (the pelvic nerve).

- The pelvic nerve acts to contract the detrusor muscle, and stimulate micturition.

Although it is non-functional post childhood, the bladder stretch reflex needs to be considered in spinal injuries (where the descending inhibition cannot reach the bladder), and in neurodegenerative diseases (where the brain is unable to generate inhibition).

Bladder Conditions

- Cystitis: Inflammation or infection of the bladder causing acute or chronic pain, discomfort, or urinary frequency or hesitancy.

- Urinary stones: Stones (calculi) may form in the kidney and travel down to the bladder. If kidney stones block urine flow to or from the bladder, they can cause severe pain.

- Bladder cancer: A tumor in the bladder is usually discovered after blood is found in the urine. Cigarette smoking and workplace chemical exposures cause most cases of bladder cancer.

- Urinary incontinence: Uncontrolled urination, which may be chronic. Urinary incontinence can result from many causes.

- Overactive bladder: The bladder muscle (detrusor) squeezes uncontrollably, causing some urine to leak out. Detrusor overactivity is a common cause of urinary incontinence.

- Hematuria: Blood in the urine. Hematuria may be harmless, or may be caused by infection or a serious condition like bladder cancer.

- Urinary retention: Urine does not exit the bladder normally due to a blockage or suppressed bladder muscle activity. The bladder may swell to hold more than a quart of urine.

- Cystocele: Weakened pelvic muscles (usually from childbirth) allow the bladder to press on the vagina. Problems with urination can result.

- Bed-wetting (nocturnal enuresis): Bed-wetting is defined as a child age 5 or older who wets the bed at least one or two times a week over at least 3 months.

- Dysuria (painful urination): Pain or discomfort during urination due to infection, irritation, or inflammation of the bladder, urethra, or external genitals.

Urethra

The urethra is the passageway between the bladder and the external part of the body, which allows urine to be excreted from the body.

Anatomy

The urethra is a thin, fibromuscular tube that begins at the lower opening of the bladder and extends through the pelvic and urogenital diaphragms to the outside of the body, called the external urethral orifice. There is a sphincter at the upper end of the urethra, which serves to close the passage and keep the urine inside the bladder.

As the passage needs to traverse the length of the penis, it is significantly longer in males than females. It is approximately 4 cm in length for females, whereas it is about 20 cm in the male body.

The male urethra consists of three parts:

- Prostatic part descends through the prostate and can dilate to change significantly in size.

- Membranous part exists between the prostate and the beginning of the penis and is the section targeted if a catheter needs to be inserted.

- Spongy part traverses the penis to end at the external urethral orifice.

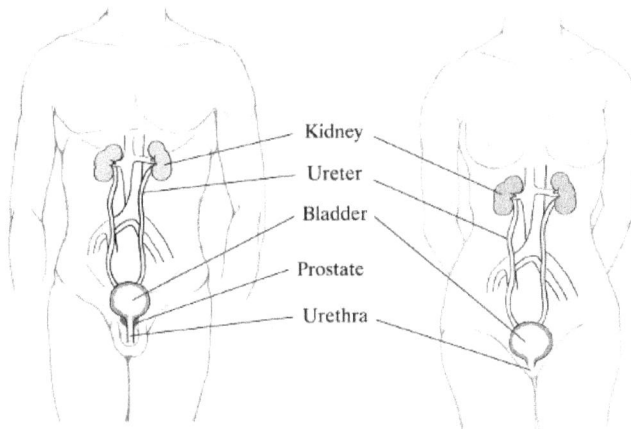

Drawing of male and female urinary tracts with the kidney, ureter, bladder, prostate (male), and urethra labeled.

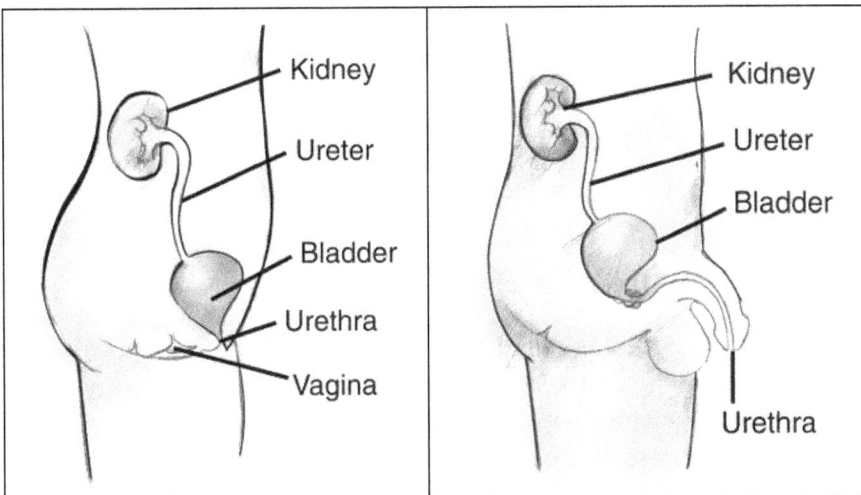

Physiology

The physiological function of the ureter is to allow the passage of urine to outside the body for excretion. When the receptors in the bladder sense that it is full, a response pathway occurs to allow the sphincter to open and the urine to pass into the urethra. This process involves both voluntary and involuntary controls, which allows individuals to dictate when they urinate unless the bladder becomes overfilled when it may occur spontaneously.

The urethra serves an additional purpose in men, as it is also utilized as a passageway for semen when a man ejaculates. Similarly, this involves a complex pathway that releases the semen from the ductus deferens for ejaculation.

Male Urethra

The male urethra is a conduit spread over the genital & urinary systems. This musculomembranous canal is running from the bladder to the end of the penis and carries the sperm but also drains the urine from the bladder outwards. A circular muscle, the sphincter, located at its initial portion, control the urination. The urethra is much longer in men VS women (15-20 cm VS 3-4 cm).

The urethra is described based on its three aspects: its embryological aspect (localization of urethra's portions), its anatomic appearance (nature of surrounding elements) and its functional aspect (degree of mobility).

Embryological and anatomical Appearance: Localization of the Urethra's Portions

The male urethra can be divided into the posterior and anterior portions

1. Posterior Urethra

   It includes the pre-prostatic urethra, the prostatic urethra and the membranous urethra.

- Pre-prostatic Urethra (also called Intra-mural)

  This uppermost segment is located in the bladder neck and surrounded by the smooth sphincter.

- Prostatic Urethra (Intra Pelvic)

  This segment begins at the bladder neck and goes through the prostate over 2 to 3 cm, keeping an almost vertical direction. Next, it joins the genital tracts and the pelvic floor (covered by the urogenital diaphragm). This portion of the urethra can be compressed in case of prostatic disease . This is the exact area where the ejaculatory ducts take form, nearby the "verumontanum", a landmark used

in classification of several urethral developmental disorders. The posterior surface of the urethra is punctuated by the prostate glands orifices.

- The Membranous Urethra

It is a short portion of 1 to 2 cm going through the pelvic floor obliquely forwards
and downwards. It is surrounded by the external sphincter (ring consisting of striated muscle fibers).

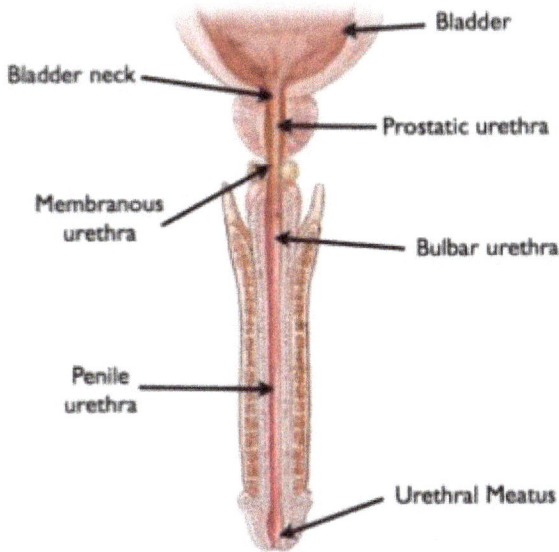

2. Anterior Urethra (Spongy Urethra)

Located in the corpus spongiosum (spongy tissues within the penis), the anterior urethra is the urethra's lowermost and longest section. It consists of the
urethra perineum, the penile urethra and ends with the urethral meatus.

- Urethra Perineal

It extends up to the root of the penis through an oblique path. It includes the
bulbar urethra located in the perineum area. It is a wide and angled zone setting
the junction between the membranous urethra and the penile urethra. It hosts
the excretory ducts of Cowper's glands producing the lubricator fluid emitted
prior the ejaculation.

- Penile Urethra

This portion of 10 to 12 cm long and a few mm in diameter goes throughout
the penis and is formed by the spongy urethra ensuring the erectile function. It
receives the secretions produced by the glands of Littre, guaranteeing a proper
lubrication of the urethra.

- Urethral Meatus

   It consists of the opening of the urethra at the penis extremity. This area is the least expandable part of the urethra canal and, in general, the narrower (less than 1 cm wide).

   The anterior urethra is closely connected to the erectile structures: its path goes through the corpora cavernosa gutter and is surrounded by the spongy body.

Functional Aspect: Mobile and fixed Portions of the Urethra

Functionally, the urethra consists of a fixed portion and a mobile portion.

- The fixed portion comprises the posterior urethra and urethra perineal (part of the anterior urethra). Its fixity is ensured by the prostate, the pelvic floor and the suspensory ligament of the penis whose elastic structure clears the angle between fixed and moving parts of the organ.
- The mobile part comprises the penile urethra.

When the penis is not erected, the urethra looks alike a reverse-S path.

Female Urethra

The female urethra is shorter than the male urethra. The female urethra is a simpler structure than the male urethra because it carries only urine.

Anatomy of Female Urethra

The female urethra which is shorter than the male urethra measures around 4 cm in length extending from the internal urethral opening in the urinary bladder to the external urinary opening situated just above the vaginal orifice. It is a narrow membranous canal placed behind the pubic symphysis (the junction between the two pubic bones) and embedded in the front wall of the vagina. It travels obliquely downward and frontward and is slightly curved with the concave part directing forward.

The diameter of the female urethra when not dilated is roughly 6 mm. During its course, it perforates a thick tissue called urogenital diaphragm. Its external orifice lies between the clitoris and the vaginal opening. Its lining membrane has longitudinal folds and one of these folds called the urethral crest is placed along the floor of the urethral canal. Several small urethral glands open into the walls of the urethra.

Structure of Female Urethra

Structurally the female urethra consists of three layers, muscular, erectile and mucous coats. The muscular coat is continuous with that of the urinary bladder and extends throughout its length. It also has a band of circular muscle fibers that act as sphincter urethrae like in a male urethra controlling the flow of urine depending on whether the bladder is full or empty. The thin layer consisting of spongy erectile tissue has a plexus of large veins intermixed with muscle fibers lies immediately beneath the mucous coat. The mucous coat is lined by striated squamous epithelium and is continuous with the bladder internally and vagina externally.

Conditions Affecting Female Urethra

Like in a male urethra, several conditions affect the female urethra. However, there is a distinct difference between the two. The female urethra being very short is more prone for urinary tract infections than in the male urethra.

Cystitis: This is a condition where the bladder and with is the urethra gets inflamed due to bacterial infections and is more common in females than in males due to the short length and susceptible for all bacteria to ascend to the bladder very easily. Every woman must have had at least one attack of cystitis during her lifetime and most of them have recurrent attacks. Pain and frequency of urination are the common symptoms and treatment lies in selecting appropriate antibiotics depending on which bacteria have invaded.

Urinary incontinence: This condition as the name implies is characterized by inability to control passage of urine resulting in involuntary passage without the knowledge of the patient. Urinary incontinence is more common in females, the cause for this condition is due to loss of control at the sphincter level at the neck of the bladder. This condition is of four types namely stress, urge, overflow and total incontinence depending on the cause. Treatment depends on the cause.

Urinary retention: Though more common in males females also suffer from this condition when there is partial or total inability to pass urine resulting in retention of urine in the bladder. Anything that exerts pressure on the urethra may lead to partial or total retention of urine. In women pressure on the bladder during early pregnancy may lead to transient retention, which disappears as soon as the pressure is released after delivery.

It is therefore essential to keep the bladder empty during labor and catheterization is a must if the bladder is full before first stage of labor. Even constipation in women may lead to pressure of the rectum on the bladder leading to retention. Stones in the bladder neck and tumors in the bladder can also lead to retention of urine. Treatment depends on the cause. However, whatever be the reason catheterization to empty the bladder is a must to prevent back flush and damage to kidneys.

Urinary tract infections: Most common in women due to easy accessibility of infections. The common infection is due to E.Coli infection. Treatment depends on the bacteria isolated on culture and sensitivity test of a urine sample. Commonly known as UTI, this infection is so common in females that women suffer 50 times more than men.

Tumors in the bladder: These can be benign or malignant. The urethra might also get involved. Early detection and remedial action can prevent invasion into the urethra and other organs nearby.

Complete disruption of female urethra: This is of rare occurrence. That may result from complete rupture of the bladder neck with laceration of the anterior vaginal wall secondary to pelvic trauma with a blunt instrument. Accidental rupture of the female urethra though uncommon may occur due to blunt accidents to the vulva.

Sexually Transmitted diseases: The female urethra like in males bears the brunt of sexually transmitted conditions like gonococcal urethritis, syphylitic urethritis and vaginal warts. Treatment depends on the cause. Post medical treatment may need dilatation of the urethra with urethral sounds if there is a stricture.

References

- Kidney, human-body-maps: healthline.com, Retrieved 23 April 2018

- Functions-of-the-kidney-6704: interactive-biology.com, Retrieved 13 June 2018

- What-is-the-Urethra, health: news-medical.net, Retrieved 27 May 2018

- Male-urethra-anatomy, diagnostic-urethra, urethral-stenosis: laser-prostate-robot.co.uk, Retrieved 21 April 2018

- Female-urethra-the-conduit-for-urinary-discharge, urinary-system, know-your-body: desimd.com, Retrieved 11 May 2018

9

Gastrointestinal System

The gastrointestinal system is an organ system, which is responsible for the digestion of food, extraction and absorption of energy and the expulsion of waste matter in the form of feces. This chapter delves into the major organs of the gastro intestinal system.

Gastrointestinal Tract

The gastrointestinal tract (GIT) consists of a hollow muscular tube starting from the oral cavity, where food enters the mouth, continuing through the pharynx, oesophagus, stomach and intestines to the rectum and anus, where food is expelled. There are various accessory organs that assist the tract by secreting enzymes to help break down food into its component nutrients. Thus the salivary glands, liver, pancreas and gall bladder have important functions in the digestive system. Food is propelled along the length of the GIT by peristaltic movements of the muscular walls.

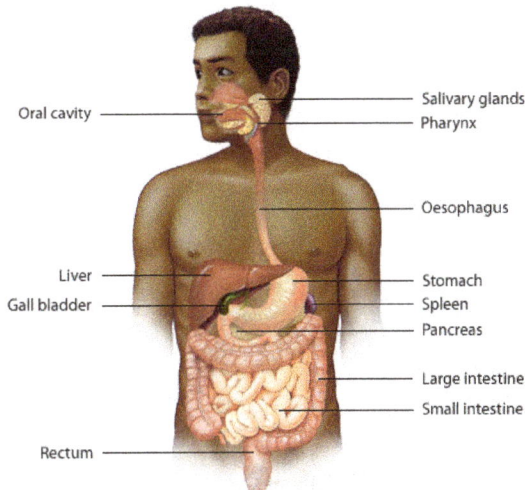

The primary purpose of the gastrointestinal tract is to break food down into nutrients, which can be absorbed into the body to provide energy. First food must be ingested into the mouth to be mechanically processed and moistened. Secondly, digestion occurs mainly in the stomach and small intestine where proteins, fats and carbohydrates are chemically broken down into their basic building blocks. Smaller molecules are then absorbed across the epithelium of the small intestine and subsequently enter the circulation. The large intestine plays a key role in reabsorbing excess water. Finally, undigested material and secreted waste products are excreted from the body via defecation (passing of faeces).

In the case of gastrointestinal disease or disorders, these functions of the gastrointestinal tract are not achieved successfully. Patients may develop symptoms of nausea, vomiting, diarrhoea, malabsorption, constipation or obstruction. Gastrointestinal problems are very common and most people will have experienced some of the above symptoms several times throughout their lives.

Basic Structure

The gastrointestinal tract is a muscular tube lined by a special layer of cells, called epithelium. The contents of the tube are considered external to the body and are in continuity with the outside world at the mouth and the anus. Although each section of the tract has specialised functions, the entire tract has a similar basic structure with regional variations.

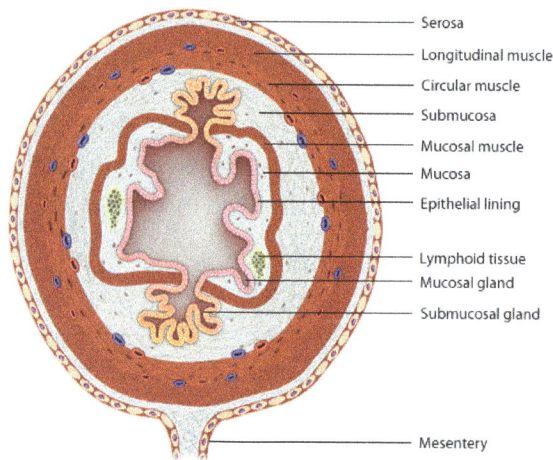

The wall is divided into four layers as follows:

Mucosa

The innermost layer of the digestive tract has specialised epithelial cells supported by an underlying connective tissue layer called the lamina propria. The lamina propria contains blood vessels, nerves, lymphoid tissue and glands that support the mucosa. Depending on its function, the epithelium may be simple (a single layer) or stratified (multiple layers).

Areas such as the mouth and oesophagus are covered by a stratified squamous (flat) epithelium so they can survive the wear and tear of passing food. Simple columnar (tall) or glandular epithelium lines the stomach and intestines to aid secretion and absorption. The inner lining is constantly shed and replaced, making it one of the most rapidly dividing areas of the body. Beneath the lamina propria is the muscularis mucosa. This comprises layers of smooth muscle which can contract to change the shape of the lumen.

Submucosa

The submucosa surrounds the muscularis mucosa and consists of fat, fibrous connective tissue and larger vessels and nerves. At its outer margin there is a specialized nerve plexus called the submucosal plexus or Meissner plexus. This supplies the mucosa and submucosa.

Muscularis Externa

This smooth muscle layer has inner circular and outer longitudinal layers of muscle fibres separated by the myenteric plexus or Auerbach plexus. Neural innervations control the contraction of these muscles and hence the mechanical breakdown and peristalsis of the food within the lumen.

Serosa/Mesentery

The outer layer of the GIT is formed by fat and another layer of epithelial cells called mesothelium.

Functions of Gastrointestinal Tract

There are three main functions of the gastrointestinal tract, including transportation, digestion, and absorption of food. The mucosal integrity of the gastrointestinal tract and the functioning of its accessory organs are vital in maintaining the health of your patient. Components of the gastrointestinal system include the mouth, esophagus, stomach, small intestine, and large intestine. The gastrointestinal tract's accessory organs include the liver, pancreas, and gallbladder.

The mouth functions to break down food into smaller parts. The esophagus is the tube that allows the passage of the food bolus from the mouth to the stomach. It plays no part in the digestive process.

The stomach functions to store, churn, and puree food into a substance known as chime. Gastric juices are secreted by the cells of the stomach, contributing to chemical digestion.

The small intestine extends from the pylorus to the ileocecal valve. The small intestine is composed of the duodenum, jejunum, and ileum. The primary function of the small intestine is the absorption of vitamins and nutrients, including electrolytes, iron, carbohydrates, proteins, and fats. Most digestion of nutrients happens here.

The large intestine extends from the terminal ileum at the ileocecal valve to the rectum. At the terminal ileum, the large intestine becomes the ascending colon, the transverse colon, and then the descending colon. Following the descending colon is the sigmoid colon and the rectum. The main function of the large intestine is water absorption. Typically, the large intestine absorbs about one and one-half liters of water per day. It can, however, absorb up to six liters.

The gallbladder is a pear-shaped, sac-like organ attached to the liver that serves as a storage facility for bile. When a large or fatty meal is consumed, nerve and chemical signals (release of the enzyme CCK) cause the gallbladder to contract. This contraction releases bile into the digestive system.

The liver is a very large organ located in the upper right abdomen. Blood supply to the liver arises from both the portal vein and hepatic artery. Nearly one-quarter of our cardiac output is delivered through the liver per minute, most of which travels through the portal vein. The blood is filtered through the liver, which destroy debris and unwanted organisms.

The pancreas is both an endocrine and exocrine gland. The exocrine function of the pancreas is mainly digestive in nature and involves the secretion of pancreatic enzymes and bicarbonate.

Upper Gastrointestinal Tract

The upper gastrointestinal tract consists of the esophagus, stomach, and duodenum. The exact demarcation between upper and lower can vary. Upon gross dissection, the duodenum may appear to be a unified organ, but it is often divided into two parts based upon function, arterial supply, or embryology.

The functions of the upper gastrointestinal tract include transport of the swallowed food bolus, enzymatic digestion, and absorption of nutrients, in addition to protective barrier function against the external environment. The morphologic appearance of the different sections of the upper digestive tract reflects the primary function of each segment and is variable between the species.

The upper gastrointestinal tract includes the:

Esophagus

The esophagus is a long, thin, and muscular tube that connects the pharynx (throat) to the stomach. It forms an important piece of the gastrointestinal tract and functions as the conduit for food and liquids that have been swallowed into the pharynx to reach the stomach.

The esophagus is about 9-10 inches (25 centimeters) long and less than an inch (2 centimeters) in diameter when relaxed. It is located just posterior to the trachea in the neck and thoracic regions of the body and passes through the esophageal hiatus of the diaphragm on its way to the stomach.

At the superior end of the esophagus is the upper esophageal sphincter that keeps the esophagus closed where it meets the pharynx. The upper esophageal sphincter opens only during the process of swallowing to permit food to pass into the esophagus. At the inferior end of the esophagus, the lower esophageal sphincter opens for the purpose of permitting food to pass from the esophagus into the stomach. Stomach acid and chyme (partially digested food) is normally prevented from entering the esophagus, thanks to the lower esophageal sphincter. If this sphincter weakens, however, acidic chyme may return to the esophagus in a condition known as acid reflux. Acid reflux can cause damage to the esophageal lining and result in a burning sensation known as heartburn. If these symptoms occur with enough frequency, they are known as GERD (gastroesophageal reflux disease).

Like the rest of the gastrointestinal tract, the esophagus is made of four distinct tissue layers.

- The mucosa layer forms the inner lining of the esophagus and is the only tissue layer that has direct contact with substances passing through the esophagus. Non-keratinized stratified squamous epithelial tissue makes up the majority of the mucosa layer and provides protection to the esophagus from rough food particles and acid from the nearby stomach. Mucous glands in the mucosa produce mucus to lubricate the esophagus and help shield the mucosa from stomach acid.

- Deep to the mucosa is the submucosa layer that contains connective tissue and provides blood and nerve supply to the mucosa and other tissues of the esophagus.

- Surrounding the submucosa is the muscularis layer that allows the esophagus to contract and expand to move substances. Skeletal muscle is mostly found in the superior region of the esophagus to aid in the swallowing reflex while smooth muscle in the inferior esophagus pushes substances toward the stomach via peristalsis.

- Finally, the adventitia layer forms an outer covering of loose connective tissue around the esophagus and attaches it loosely to the surrounding organs.

The esophagus is involved in the processes of swallowing and peristalsis to move substances from the mouth to the stomach. The swallowing food begins in the mouth and continues with the contraction of skeletal muscles in the pharynx and esophagus. The upper esophageal sphincter dilates to permit the swallowed substance to enter the esophagus. From this point, waves of muscle contraction called peristalsis move food toward the stomach. In peristalsis, regions of the esophagus closer to the stomach open to permit food to pass through while the region just above the food contracts to push the food onward. Peristalsis works so well that food can be swallowed even while the body is lying down, upside down, or even in zero-gravity.

A final function of the esophagus is its participation in the vomiting reflex to void the contents of the stomach. Peristalsis is reversed in the esophagus during vomiting to forcefully remove toxic or pathogen-laden food from the body.

Esophagus Conditions

- Heartburn: An incompletely closed LES allows acidic stomach contents to back up (reflux) into the esophagus. Reflux can cause heartburn, cough or hoarseness, or no symptoms at all.

- Gastroesophageal reflux disease (GERD): When reflux occurs frequently or is bothersome, it's called gastroesophageal reflux disease (GERD).

- Esophagitis: Inflammation of the esophagus. Esophagitis can be due to irritation (as from reflux or radiation treatment) or infection.

- Barrett's esophagus: Regular reflux of stomach acid irritates the esophagus, which may cause the lower part to change its structure. Very infrequently, Barrett's esophagus progresses to esophageal cancer.

- Esophageal ulcer: An erosion in an area of the lining of the esophagus. This is often caused by chronic reflux.

- Esophageal stricture: A narrowing of the esophagus. Chronic irritation from reflux is the usual cause of esophageal strictures.

- Achalasia: A rare disease in which the lower esophageal sphincter does not relax properly. Difficulty swallowing and regurgitation of food are symptoms.

- Esophageal cancer: Although serious, cancer of the esophagus is uncommon. Risk factors for esophageal cancer include smoking, heavy drinking, and chronic reflux.

- Mallory-Weiss tear: Vomiting or retching creates a tear in the lining of the esophagus. The esophagus bleeds into the stomach, often followed by vomiting blood.

- Esophageal varices: In people with cirrhosis, veins in the esophagus may become engorged and bulge. Called varices, these veins are vulnerable to life-threatening bleeding.

- Esophageal ring (Schatzki's ring): A common, benign accumulation of tissue in a ring around the low end of the esophagus. Schatzki's rings usually cause no symptoms, but may cause difficulty swallowing.

- Esophageal web: An accumulation of tissue (similar to an esophageal ring) that usually occurs in the upper esophagus. Like rings, esophageal webs usually cause no symptoms.

- Plummer-Vinson syndrome: A condition including chronic iron-deficient anemia, esophageal webs, and difficulty swallowing. Iron replacement and dilation of esophageal webs are treatments.

- Esophageal stricture: A narrowing of the esophagus, from a variety of causes, which, if narrow enough, may lead to difficult swallowing.

Stomach

The stomach is a muscular organ that is found in our upper abdomen. If we were to locate it on our bodies, it can be found on our left side just below the ribs. In simple terms, the stomach is a kind of digestive sac. It is a continuation of the esophagus and receives our churned food from it. Therefore, the stomach serves as a kind of connection between the esophagus and the small intestine, and is a definite pit stop along our alimentary canal. Muscular sphincters, which are similar to valves, allow some separation between these organs.

The stomach's functions benefit from several morphological attributes. The stomach is able to secrete enzymes and acid from its cells, which enables it to perform its digestive functions. With its muscular lining, the stomach is able to engage in peristalsis (in other words, to form the ripples that propel the digested food forward) and in the general "churning" of food. Likewise, the abundant muscular tissue of the stomach has ridges in its linings called rugae. These increase the surface area of the stomach and facilitate its functions.

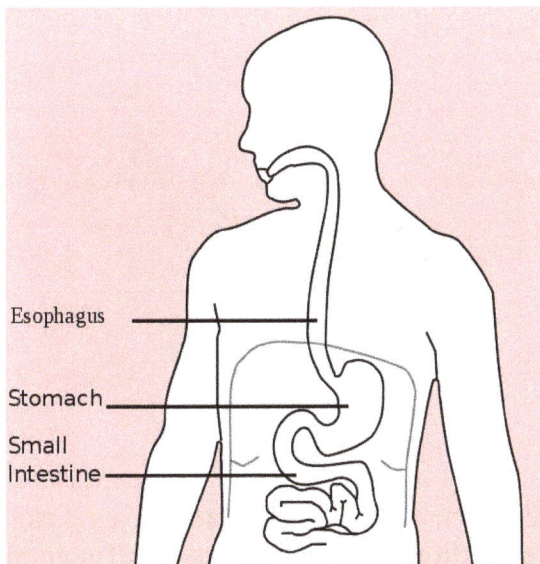

Esophagus, stomach, and intestinal regions of the human body

Functions of the Stomach

The stomach is first and foremost a principal site of digestion. In fact, it is the first site of actual protein digestion. While sugars can begin to be lightly digested by salivary enzymes in the mouth, protein degradation will not occur until the food bolus reaches the stomach. This breakdown is carried out by the stomach's *pepsin* enzyme. The stomach's roles can essentially be distilled down to three functions.

Much like an elastic bag, the stomach will provide a place for varied amounts of swallowed food to rest and digest in. Hence, the stomach is a storage site. The stomach will also introduce our swallowed food to essential acids. The cells in the stomach's lining will excrete a strong acidic mixture of hydrochloric acid, sodium chloride, and potassium chloride. This gastric acid, or colloquially known as gastric "juice," will work to break down the bonds within the food particles at the molecular level. Pepsin enzyme will have the unique role of breaking the strong peptide bonds that hold the proteins in our food together, further preparing the food for the nutrient absorption that takes place in the small (mainly) and large intestines. This brings us to the third task the stomach has, which is to send off the churned watery mixture to the small intestine for further digestion and absorption. It takes about three hours for this to occur once the food is a liquid mix.

The stomach's main roles:

1. Food storage.

2. Acidic breakdown of swallowed food.

3. Sends mixture on to the next phase in the small intestine.

Structure of the Stomach

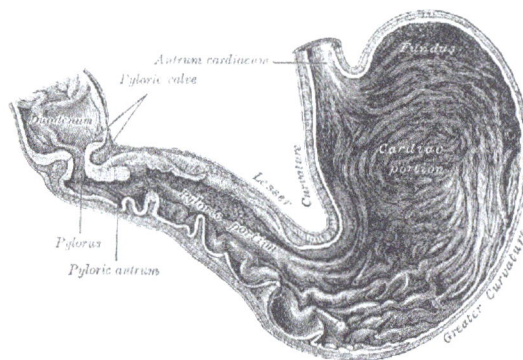

Different regions of the stomach

Although we have briefly discussed the location and physical traits of the stomach, it is important to detail the structure of the stomach, as well. The stomach begins at the lower esophageal sphincter that discerns the cut-off point of the esophagus. The stomach itself is very muscular. When the muscularis externa layers are dissected, one can visualize three distinct layers coined the longitudinal, circular, and oblique layers. The first region of the stomach is called the cardia. It is the layer closest to the esophagus and it contains cardiac glands that secrete mucus. Mucus protects the delicate epithelial lining of many tissues in the human body. This region is followed by the fundus, which is the superior arch of the stomach. Importantly, the fundus has the special function of containing gastric glands that release a cocktail of gastric juices. This region is followed by the body of the stomach, which is coated with rugae and is the largest region. Rugae, in turn, help facilitate digestion by increasing the site's surface area. Finally, this section is followed by the pylorus region, which is closest to the exit into the duodenum of the small intestine and is pinched off by the pyloric sphincter.

Four regions of the stomach:

- Cardiac

- Fundus

- Body

- Pylorus

Stomach Conditions

- Gastroesophageal reflux: Stomach contents, including acid, can travel backward up the esophagus. There may be no symptoms, or reflux may cause heartburn or coughing.

- Gastroesophageal reflux disease (GERD): When symptoms of reflux become bothersome or occur frequently, they're called GERD. Infrequently, GERD can cause serious problems of the esophagus.

- Dyspepsia: Another name for stomach upset or indigestion. Dyspepsia may be caused by almost any benign or serious condition that affects the stomach.

- Gastric ulcer (stomach ulcer): An erosion in the lining of the stomach, often causing pain and/or bleeding. Gastric ulcers are most often caused by NSAIDs or *H. pylori* infection.

- Peptic ulcer disease: Doctors consider ulcers in either the stomach or the duodenum (the first part of the small intestine) peptic ulcer disease.

- Gastritis: Inflammation of the stomach, often causing nausea and/or pain. Gastritis can be caused by alcohol, certain medications, *H. pylori*infection, or other factors.

- Stomach cancer: Gastric cancer is an uncommon form of cancer in the U.S. Adenocarcinoma and lymphoma make up most of the cases of stomach cancer.

- Zollinger-Ellison syndrome (ZES): One or more tumors that secrete hormones that lead to increased acid production. Severe GERD and peptic ulcer disease result from this rare disorder.

- Gastric varices: In people with severe liver disease, veins in the stomach may swell and bulge under increased pressure. Called varices, these veins are at high risk for bleeding, although less so than esophageal varices are.

- Stomach bleeding: Gastritis, ulcers, or gastric cancers may bleed. Seeing blood or black material in vomit or stool is usually a medical emergency.

- Gastroparesis (delayed gastric emptying): Nerve damage from diabetes or other conditions may impair the stomach's muscle contractions. Nausea and vomiting are the usual symptoms.

Duodenum

The duodenum is the first and shortest segment of the small intestine. It receives partially digested food (known as *chyme*) from the stomach and plays a vital role in the chemical digestion of chyme in preparation for absorption in the small intestine. Many chemical secretions from the pancreas, liver and gallbladder mix with the chyme in the duodenum to facilitate chemical digestion.

Structure and Location

The duodenum is located just below the stomach, and its first part is quite close to the liver and the pancreas. It is located between the stomach and the jejunum. Anatomically, this small organ is divided into four parts or segments, which are known as the superior, descending, horizontal, and ascending duodenum.

It begins with the duodenal bulb, which is located next to the stomach. The first or superior part of the duodenum is nothing but a continuation of the duodenal end of the pylorus. The second or the descending part is the portion, where the duodenum begins to curve or descend. This is where the common bile duct and the pancreatic duct enter the duodenum.

The third part of the duodenum is the horizontal portion. The fourth or the ascending part, connects to the diaphragm by the ligament of Treitz, and then leads to the jejunum. The duodenum ends at the duodenojejunal flexure, the point where it meets the jejunum.

Functions of the Duodenum

Its main function is to receive partially-digested food from the stomach, and then complete the process of digestion. The first part of the duodenum is more susceptible to peptic ulcers, mainly due to its exposure to the chyme that contains unneutralized stomach acids. As one of the most important parts of the digestive system, it is concerned with both digestion of food and absorption of nutrients. In fact, most chemical digestion takes place in the duodenum.

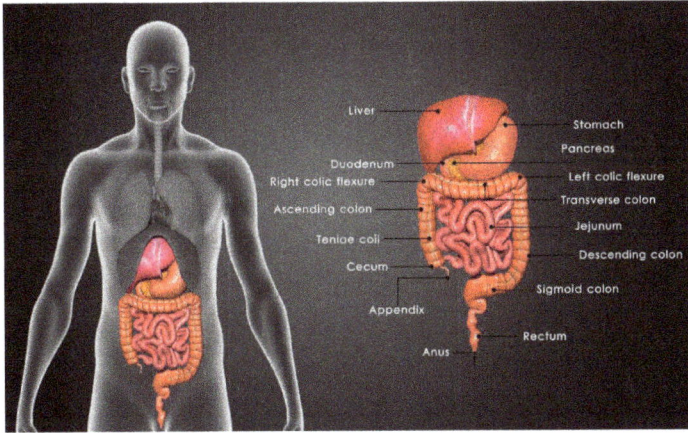

Importance in Digestion

- In the stomach, food is mixed with stomach acids, and then digested partially. This mixture of semi-digested food and stomach acids is known as chyme. The duodenum receives the acidic chyme from the stomach, and then starts breaking it down with the help of enzymes and intestinal juice. There is a small circular opening between the stomach and the duodenum, known as pylorus, which releases the chyme into the duodenum.

- The cells present in the duodenal epithelium release two hormones, known as secretin and cholecystokinin. Secretin is released in response to the presence of excess acids in the duodenum, while the secretion of cholecystokinin is stimulated by the presence of amino acids and fatty acids.

- The hormone secretin stimulates the gallbladder to release alkaline bile, while cholecystokinin induces the pancreas to secrete bicarbonates into the duodenum. The pancreas also releases several enzymes (lipase, amylase, and trypsin) into the duodenum. These secretions increase the pH of acidic chyme, and play a crucial role in digestion.

- The inner lining of the duodenum contains crypts. These crypts are responsible for increasing the surface area of the intestinal membrane, and thereby ensure better digestion. The duodenum also contains smooth muscles, which facilitate the movement of waste materials down the large intestine.

- The duodenum is responsible for regulating the rate of gastric emptying, and triggering the hunger signals. Gastric emptying refers to the emptying of the stomach, i.e., the process of food leaving the stomach and entering the duodenum. These functions are facilitated by the hormones secreted by the duodenal epithelium.

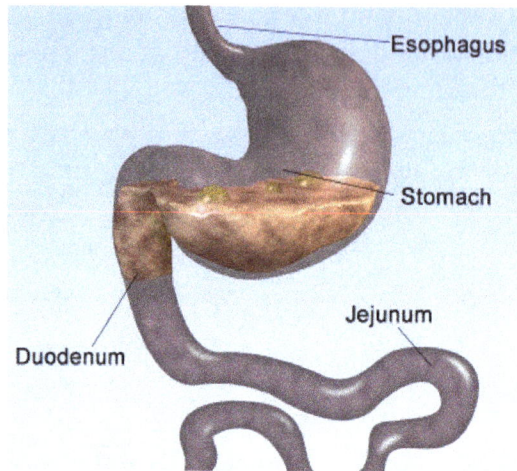

Other Functions

- Absorption of many essential nutrients takes place in the duodenum. It houses the Brunner's glands, which are responsible for producing a mucus-rich alkaline secretion that serves several important purposes. It protects the duodenum from the acidic chyme and releases urogastrone, which prevents the secretion of acid and other digestive enzymes by the parietal and chief cells of the stomach.

- The duodenum is the site of iron and calcium absorption. Other nutrients that are absorbed in the duodenum are Vitamin A and B1, amino acids, fatty acids, monoglycerides, phosphorus, and mono and disaccharides.

- The base of the duodenum, jejunum, and ileum contains the paneth cells, that secrete antibacterial enzymes to protect the intestine from microbes. These cells help prevent the excessive growth of intestinal flora.

To sum up, the duodenum is concerned with the digestion of food, absorption of nutrients, regulation of the rate of gastric emptying and hunger, and the movement of food through the intestinal tract. The gastric bypass surgery (used for treating morbid obesity) often involves the duodenum, due to its immense importance in the absorption of nutrients, and regulation of hunger and movement of food.

Lower Gastrointestinal Tract

Lower gastrointestinal tract is also known as your bowel which is approximately 25 feet long and consists of the small intestine and large intestine.

Food from the stomach passes through the pyloric valve into the small intestine, a 20-foot tube with three sections: the duodenum, the jejunum, and the ileum. The walls of the small intestine are lined with muscles that contract and relax to carry food along its path. The walls are also covered in microvilli, hairlike projections that help to absorb nutrients into the bloodstream.

When food reaches the duodenum, it begins to get broken down by digestive enzymes and bile. This process turns proteins into amino acids, fats into fatty acids, and carbohydrates into simple sugars. The digested food then moves into the jejunum, where most of its nutrients are absorbed. Vitamin B12 is absorbed in the ileum.

The material left behind—mostly water, electrolytes (such as sodium and potassium), and waste (such as fiber and dead cells)—moves into the cecum, the first part of the colon (also called the large intestine). It then passes through the ascending, transverse, and descending colons; the sigmoid colon; and the rectum.

No nutrients are absorbed by the colon. Its job is to remove excess water from the intestinal waste and return it to the bloodstream. Thus, as the material moves along the colon, it slowly dries out and forms a more solid substance called stool.

Waste usually spends a day or two in the colon before it is expelled from the body. When stool moves into the rectum, it stretches the walls of the rectum, which signals the need for a bowel movement. The stool then moves into the anal canal. At the end of the anal canal is the anal sphincter, which is a muscle that usually remains closed; however, it opens to allow stool to pass out of the body.

Small Intestine

The small intestine is a long, highly convoluted tube in the digestive system that absorbs about 90% of the nutrients from the food we eat. It is given the name "small intestine" because it is only 1 inch in diameter, making it less than half the diameter of the large intestine. The small intestine is, however, about twice the length of the large intestine and usually measures about 10 feet in length.

The small intestine winds throughout the abdominal cavity inferior to the stomach. Its many folds help it to pack all 10 feet of its length into such a small body cavity.

A thin membrane known as the mesentery extends from the posterior body wall of the abdominal cavity to surround the small intestine and anchor it in place. Blood vessels, nerves, and lymphatic vessels pass through the mesentery to support the tissues of the small intestine and transport nutrients from food in the intestines to the rest of the body.

The small intestine can be divided into 3 major regions:

1. The duodenum is the first section of intestine that connects to the pyloric sphincter of the stomach. It is the shortest region of the small intestine, measuring only about 10 inches in length. Partially digested food, or *chyme*,

from the stomach is mixed with bile from the liver and pancreatic juice from the pancreas to complete its digestion in the duodenum.

2. The jejunum is the middle section of the small intestine that serves as the primary site of nutrient absorption. It measures around 3 feet in length.

3. The ileum is the final section of the small intestine that empties into the large intestine via the ileocecal sphincter. The ileum is about 6 feet long and completes the absorption of nutrients that were missed in the jejunum.

Like the rest of the gastrointestinal tract, the small intestine is made up of four layers of tissue. The mucosa forms the inner layer of epithelial tissue and is specialized for the absorption of nutrients from chyme. Deep to the mucosa is the submucosa layer that provides blood vessels, lymphatic vessels, and nerves to support the mucosa on the surface. Several layers of smooth muscle tissue form the muscularis layer that contracts and moves the small intestines. Finally, the serosa forms the outermost layer of epithelial tissue that is continuous with the mesentery and surrounds the intestines.

The interior walls of the small intestine are tightly wrinkled into projections called circular folds that greatly increase their surface area. Microscopic examination of the mucosa reveals that the mucosal cells are organized into finger-like projections known as villi, which further increase the surface area. Each square inch of mucosa contains around 20,000 villi. The cells on the surface of the mucosa also contain finger-like projections of their cell membranes known as microvilli, which further increase the surface area of the small intestine. It is estimated that there are around 130 billion microvilli per square inch in the mucosa of the small intestine. All of these wrinkles and projections help to greatly increase the amount of contact between the cells of the mucosa and chyme to maximize the absorption of vital nutrients.

The small intestine processes around 2 gallons of food, liquids, and digestive secretions every day. To ensure that the body receives enough nutrients from its food, the small intestine mixes the chyme using smooth muscle contractions called *segmentations*. Segmentation involves the mixing of chyme about 7 to 12 times per minute within a short segment of the small intestine so that chyme in the middle of the intestine is moved outward to the intestinal wall and contacts the mucosa. In the duodenum, segmentations help to mix chyme with bile and pancreatic juice to complete the chemical digestion of the chyme into its component nutrients. Villi and microvilli throughout the intestines sway back and forth during the segmentations to increase their contact with chyme and efficiently absorb nutrients.

Once nutrients have been absorbed by the mucosa, they are passed on into tiny blood vessels and lymphatic vessels in the middle of the villi to exit through the mesentery. Fatty acids enter small lymphatic vessels called lacteals that carry them back to the blood supply. All other nutrients are carried through veins to the liver, where many nutrients are stored and converted into useful energy sources.

Chyme is slowly passed through the small intestine by waves of smooth muscle contraction known as *peristalsis*. Peristalsis waves begin at the stomach and pass through the duodenum, jejunum, and finally the ileum. Each wave moves the chyme a short distance, so it takes many waves of peristalsis over several hours to move chyme to the end of the ileum.

Large Intestine

The large intestine is the final section of the gastrointestinal tract that performs the vital task of absorbing water and vitamins while converting digested food into feces. Although shorter than the small intestine in length, the large intestine is considerably thicker in diameter, thus giving it its name. The large intestine is about 5 feet (1.5 m) in length and 2.5 inches (6-7 cm) in diameter in the living body, but becomes much larger postmortem as the smooth muscle tissue of the intestinal wall relaxes.

The large intestine wraps around the border of the abdominal body cavity from the right side of the body, across the top of the abdomen, and finally down the left side.

Beginning on the right side of the abdomen, the large intestine is connected to the ilium of the small intestine via the ileocecal sphincter. From the ileocecal sphincter, the large intestine forms a sideways "T," extending both superiorly and inferiorly. The inferior region of the large intestine forms a short dead-end segment known as the cecum that terminates in the vermiform appendix. The superior region forms a hollow tube known as the ascending colon that climbs along the right side of the abdomen. Just inferior to the diaphragm, the ascending colon turns about 90 degrees toward the middle of the body at the hepatic flexure and continues across the abdomen as the transverse colon. At the left side of the abdomen, the transverse colon turns about 90 degrees at the splenic flexure and runs down the left side of the abdomen as the descending colon. At the end of the descending colon, the large intestine bends slightly medially at the sigmoid flexure to form the S-shaped sigmoid colon before straightening into the rectum. The rectum is the enlarged final segment of the large intestine that terminates at the anus.

Like the rest of the gastrointestinal canal, the large intestine is made of four tissue layers:

- The innermost layer, known as the *mucosa*, is made of simple columnar epithelial tissue. The mucosa of the large intestine is smooth, lacking the villi found in the small intestine. Many mucous glands secrete mucus into the hollow lumen of the large intestine to lubricate its surface and protect it from rough food particles.

- Surrounding the mucosa is a layer of blood vessels, nerves and connective tissue known as the *submucosa*, which supports the other layers of the large intestine.

- The *muscularis* layer surrounds the submucosa and contains many layers of visceral muscle cells that contract and move the large intestine. Continuous contraction of smooth muscle bands in the muscularis produces lumpy, pouch-like structures known as *haustra* in the large intestine.

- Finally, the *serosa* forms the outermost layer. The serosa is a thin layer of simple squamous epithelial tissue that secretes watery serous fluid to lubricate the surface of the large intestine, protecting it from friction between abdominal organs and the surrounding muscles and bones of the lower torso.

The large intestine performs the vital functions of converting food into feces, absorbing essential vitamins produced by gut bacteria, and reclaiming water from feces. A slurry of digested food, known as *chyme*, enters the large intestine from the small intestine via the ileocecal sphincter. Chyme passes through the cecum where it is mixed with beneficial bacteria that have colonized the large intestine throughout a person's lifetime. The chyme is then slowly moved from one haustra to the next through the four regions of the colon. Most of the movement of chyme is achieved by slow waves of peristalsis over a period of several hours, but the colon can also be emptied quickly by stronger waves of mass peristalsis following a large meal.

While chyme moves through the large intestine, bacteria digest substances in the chyme that are not digestible by the human digestive system. Bacterial fermentation converts the chyme into feces and releases vitamins including vitamins K, B1, B2, B6, B12, and biotin. Vitamin K is almost exclusively produced by the gut bacteria and is essential in the proper clotting of blood. Gases such as carbon dioxide and methane are also produced as a byproduct of bacterial fermentation and lead to flatulence, or gas passed through the anus.

The absorption of water by the large intestine not only helps to condense and solidify feces, but also allows the body to retain water to be used in other metabolic processes. Ions and nutrients released by gut bacteria and dissolved in water are also absorbed in the large intestine and used by the body for metabolism. The dried, condensed fecal matter is finally stored in the rectum and sigmoid colon until it can be eliminated from the body through the process of defecation.

Intestine Conditions

- Stomach flu (enteritis): Inflammation of the small intestine. Infections (from viruses, bacteria, or parasites) are the common cause.

- Small intestine cancer: Rarely, cancer may affect the small intestine. There are multiple types of small intestine cancer, causing about 1,100 deaths each year.

- Celiac disease: An "allergy" to gluten (a protein in most breads) causes the small intestine not to absorb nutrients properly. Abdominal pain and weight loss are usual symptoms.

- Carcinoid tumor: A benign or malignant growth in the small intestine. Diarrhea and skin flushing are the most common symptoms.

- Intestinal obstruction: A section of either the small or large bowel can become blocked or twisted or just stop working. Belly distension, pain, constipation, and vomiting are symptoms.

- Colitis: Inflammation of the colon. Inflammatory bowel disease or infections are the most common causes.

- Diverticulosis: Small weak areas in the colon's muscular wall allow the colon's lining to protrude through, forming tiny pouches called diverticuli. Diverticuli usually cause no problems, but can bleed or become inflamed.

- Diverticulitis: When diverticuli become inflamed or infected, diverticulitis results. Abdominal pain and constipation are common symptoms.

- Colon bleeding (hemorrhage): Multiple potential colon problems can cause bleeding. Rapid bleeding is visible in the stool, but very slow bleeding might not be.

- Inflammatory bowel disease: A name for either Crohn's disease or ulcerative colitis. Both conditions can cause colon inflammation (colitis).

- Crohn's disease: An inflammatory condition that usually affects the colon and intestines. Abdominal pain and diarrhea (which may be bloody) are symptoms.

- Ulcerative colitis: An inflammatory condition that usually affects the colon and rectum. Like Crohn's disease, bloody diarrhea is a common symptom of ulcerative colitis.

- Diarrhea: Stools that are frequent, loose, or watery are commonly called diarrhea. Most diarrhea is due to self-limited, mild infections of the colon or small intestine.

- Salmonellosis: Salmonella bacteria can contaminate food and infect the intestine. Salmonella causes diarrhea and stomach cramps, which usually resolve without treatment.

- Shigellosis: Shigella bacteria can contaminate food and infect the intestine. Symptoms include fever, stomach cramps, and diarrhea, which may be bloody.

- Traveler's diarrhea: Many different bacteria commonly contaminate water or food in developing countries. Loose stools, sometimes with nausea and fever, are symptoms.

- Colon polyps: Polyps are growths inside the colon. Colon cancer can often develop in these tumors after many years.

- Colon cancer: Cancer of the colon affects more than 100,000 Americans each year. Most colon cancer is preventable through regular screening.

- Rectal cancer: Colon and rectal cancer are similar in prognosis and treatment. Doctors often consider them together as colorectal cancer.

- Constipation: When bowel movements are infrequent or difficult.

- Irritable bowel syndrome (IBS): Irritable bowel syndrome, also known as IBS, is an intestinal disorder that causes irritable abdominal pain or discomfort, cramping or bloating, and diarrhea or constipation.

- Rectal prolapse: Part or all of the wall of the rectum can move out of position, sometimes coming out of the anus, when straining during a bowel movement.

- Intussusception: Occurring mostly in children, the small intestine can collapse into itself like a telescope. It can become life-threatening if not treated.

References

- Gastrointestinal-system, clinical-insights: rn.com, Retrieved 21 July 2018

- Overview-of-the-digestive-system: lumenlearning.com, Retrieved 29 June 2018

- Upper-digestive-tract, medicine-and-dentistry: sciencedirect.com, Retrieved 19 July 2018

- Duodenum-function: bodytomy.com, Retrieved 20 May 2018

- Large-intestine, digestive, anatomy: innerbody.com, Retrieved 21 April 2018

10

Reproductive System

The reproductive system consists of sex organs that facilitate sexual reproduction in humans. This chapter provides an overview of the human reproductive system and includes valuable insights into the functions of the male and female reproductive system.

Human reproductive system is an organ system by which humans reproduce and bear live offspring. The essential features of human reproduction are:

(1) Liberation of an ovum, or egg, at a specific time in the reproductive cycle,

(2) Internal fertilization of the ovum by spermatozoa, or sperm cells,

(3) Transport of the fertilized ovum to the uterus, or womb,

(4) Implantation of the blastocyst, the early embryo developed from the fertilized ovum, in the wall of the uterus,

(5) Formation of a placenta and maintenance of the unborn child during the entire period of gestation,

(6) Birth of the child and expulsion of the placenta, and

(7) Suckling and care of the child, with an eventual return of the maternal organs to virtually their original state.

For this biological process to be carried out, certain organs and structures are required in both the male and the female. The source of the ova (the female germ cells) is the female ovary; that of spermatozoa (the male germ cells) is the testis. In females, the two ovaries are situated in the pelvic cavity; in males, the two testes are enveloped in a sac of skin, the scrotum, lying below and outside the abdomen. Besides producing the germ cells, or gametes, the ovaries and testes are the source of hormones that cause full development of secondary sexual characteristics and also the proper functioning of the reproductive tracts. These tracts comprise the fallopian tubes, the uterus, the vagina, and associated structures in females and the penis, the sperm channels (epididymis, ductus deferens, and ejaculatory ducts), and other related structures and glands in males. The function of the fallopian tube is to convey an ovum, which is fertilized in the tube, to the uterus, where gestation (development before birth) takes place. The function of the male ducts is to convey spermatozoa from the testis, to store them, and, when ejaculation occurs, to eject them with secretions from the male glands through the penis.

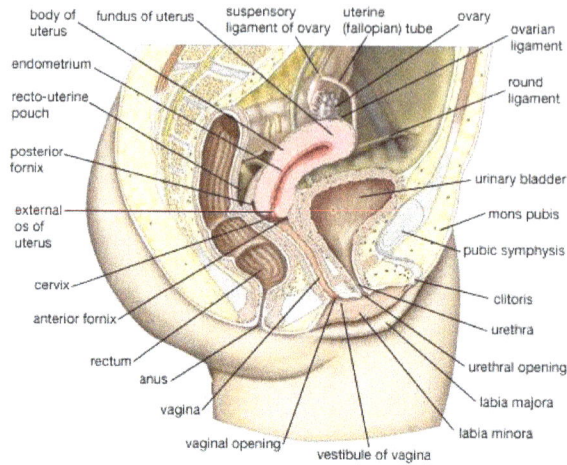

Organs of the female reproductive system.

At copulation, or sexual intercourse, the erect penis is inserted into the vagina, and spermatozoa contained in the seminal fluid (semen) are ejaculated into the female genital tract. Spermatozoa then pass from the vagina through the uterus to the fallopian tube to fertilize the ovum in the outer part of the tube. Females exhibit a periodicity in the activity of their ovaries and uterus, which starts at puberty and ends at the menopause. The periodicity is manifested by menstruation at intervals of about 28 days; important changes occur in the ovaries and uterus during each reproductive, or menstrual, cycle. Periodicity, and subsequently menstruation, is suppressed during pregnancy and lactation.

Male Reproductive System

The male reproductive system includes the scrotum, testes, spermatic ducts, sex glands, and penis. These organs work together to produce sperm, the male gamete, and the other components of semen. These organs also work together to deliver semen out of the body and into the vagina where it can fertilize egg cells to produce offspring.

Anatomy of the Male Reproductive System

Scrotum

The scrotum is a sac-like organ made of skin and muscles that houses the testes. It is located inferior to the penis in the pubic region. The scrotum is made up of 2 side-by-side pouches with a testis located in each pouch. The smooth muscles that make up the scrotum allow it to regulate the distance between the testes and the rest of the body. When the testes become too warm to support spermatogenesis, the scrotum relaxes to move the testes away from the body's heat. Conversely, the scrotum contracts to move the testes closer to the body's core heat when temperatures drop below the ideal range for spermatogenesis.

Testes

The 2 testes, also known as testicles, are the male gonads responsible for the production of sperm and testosterone. The testes are ellipsoid glandular organs around 1.5 to 2 inches long and an inch in diameter. Each testis is found inside its own pouch on one side of the scrotum and is connected to the abdomen by a spermatic cord and cremaster muscle. The cremaster muscles contract and relax along with the scrotum to regulate the temperature of the testes. The inside of the testes is divided into small compartments known as lobules. Each lobule contains a section of seminiferous tubule lined with epithelial cells. These epithelial cells contain many stem cells that divide and form sperm cells through the process of spermatogenesis.

Epididymis

The epididymis is a sperm storage area that wraps around the superior and posterior edge of the testes. The epididymis is made up of several feet of long, thin tubules that are tightly coiled into a small mass. Sperm produced in the testes moves into the epididymis to mature before being passed on through the male reproductive organs. The length of the epididymis delays the release of the sperm and allows them time to mature.

Spermatic Cords and Ductus Deferens

Within the scrotum, a pair of spermatic cords connects the testes to the abdominal cavity. The spermatic cords contain the ductus deferens along with nerves, veins, arteries, and lymphatic vessels that support the function of the testes.

The ductus deferens, also known as the vas deferens, is a muscular tube that carries sperm superiorly from the epididymis into the abdominal cavity to the ejaculatory duct. The ductus deferens is wider in diameter than the epididymis and uses its internal space to store mature sperm. The smooth muscles of the walls of the ductus deferens are used to move sperm towards the ejaculatory duct through peristalsis.

Seminal Vesicles

The seminal vesicles are a pair of lumpy exocrine glands that store and produce some of the liquid portion of semen. The seminal vesicles are about 2 inches in length and located posterior to the urinary bladder and anterior to the rectum. The liquid produced by the seminal vesicles contains proteins and mucus and has an alkaline pH to help sperm survive in the acidic environment of the vagina. The liquid also contains fructose to feed sperm cells so that they survive long enough to fertilize the oocyte.

Ejaculatory Duct

The ductus deferens passes through the prostate and joins with the urethra at a structure known as the ejaculatory duct. The ejaculatory duct contains the ducts from the seminal vesicles as well. During ejaculation, the ejaculatory duct opens and expels sperm and the secretions from the seminal vesicles into the urethra.

Urethra

Semen passes from the ejaculatory duct to the exterior of the body via the urethra, an 8 to 10 inch long muscular tube. The urethra passes through the prostate and ends at the external urethral orifice located at the tip of the penis. Urine exiting the body from the urinary bladder also passes through the urethra.

Prostate

The prostate is a walnut-sized exocrine gland that borders the inferior end of the urinary bladder and surrounds the urethra. The prostate produces a large portion of the fluid that makes up semen. This fluid is milky white in color and contains enzymes, proteins, and other chemicals to support and protect sperm during ejaculation. The prostate also contains smooth muscle tissue that can constrict to prevent the flow of urine or semen.

Unfortunately the prostate is also particularly susceptible to cancer. Thankfully, DNA health testing can tell you whether you're at higher genetic risk of developing prostate cancer due to your BRCA1 and BRCA2 genes.

Cowper's Glands

The Cowper's glands, also known as the bulbourethral glands, are a pair of pea-sized exocrine glands located inferior to the prostate and anterior to the anus. The Cowper's glands secrete a thin alkaline fluid into the urethra that lubricates the urethra and neutralizes acid from urine remaining in the urethra after urination. This fluid enters the urethra during sexual arousal prior to ejaculation to prepare the urethra for the flow of semen.

Penis

The penis is the male external sexual organ located superior to the scrotum and inferior to the umbilicus. The penis is roughly cylindrical in shape and contains the urethra and the external opening of the urethra. Large pockets of erectile tissue in the penis allow it to fill with blood and become erect. The erection of the penis causes it to increase in size and become turgid. The function of the penis is to deliver semen into the vagina during sexual intercourse. In addition to its reproductive function, the penis also allows for the excretion of urine through the urethra to the exterior of the body.

Semen

Semen is the fluid produced by males for sexual reproduction and is ejaculated out of the body during sexual intercourse. Semen contains sperm, the male reproductive gametes, along with a number of chemicals suspended in a liquid medium. The chemical composition of semen gives it a thick, sticky consistency and a slightly alkaline pH. These traits help semen to support reproduction by helping sperm to remain within the vagina after intercourse and to neutralize the acidic environment of the vagina. In healthy adult males, semen contains around 100 million sperm cells per milliliter. These sperm cells fertilize oocytes inside the female fallopian tubes.

Physiology of the Male Reproductive System

Spermatogenesis

Spermatogenesis is the process of producing sperm and takes place in the testes and epididymis of adult males. Prior to puberty, there is no spermatogenesis due to the lack of hormonal triggers. At puberty, spermatogenesis begins when luteinizing hormone (LH) and follicle stimulating hormone (FSH) are produced. LH triggers the production of testosterone by the testes while FSH triggers the maturation of germ cells. Testosterone stimulates stem cells in the testes known as spermatogonium to undergo the process of developing into spermatocytes. Each diploid spermatocyte goes through the process of meiosis I and splits into 2 haploid secondary spermatocytes. The secondary spermatocytes go through meiosis II to form 4 haploid spermatid cells. The spermatid cells then go through a process known as spermiogenesis where they grow a flagellum and develop the structures of the sperm head. After spermiogenesis, the cell is finally a sperm cell, or spermatozoa. The spermatozoa are released into the epididymis where they complete their maturation and become able to move on their own.

Fertilization

Fertilization is the process by which a sperm combines with an oocyte, or egg cell, to produce a fertilized zygote. The sperm released during ejaculation must first swim through the vagina and uterus and into the fallopian tubes where they may find an oocyte. After encountering the oocyte, sperm next have to penetrate the outer corona radiata and zona pellucida layers of the oocyte. Sperm contain enzymes in the acrosome region of the head that allow them to penetrate these layers. After penetrating the interior of the oocyte, the nuclei of these haploid cells fuse to form a diploid cell known as a zygote. The zygote cell begins cell division to form an embryo.

Functions of Male Reproductive System

The male sex organs work together to produce and release semen into the reproductive system of the female during sexual intercourse. The male reproductive system also produces sex hormones, which help a boy develop into a sexually mature man during puberty.

When a baby boy is born, he has all the parts of his reproductive system in place, but it isn't until puberty that he is able to reproduce. When puberty begins, usually between the ages of 9 and 15, the pituitary gland — which is located near the brain — secretes hormones that stimulate the testicles to produce testosterone. The production of testosterone brings about many physical changes.

Although the timing of these changes is different for every guy, the stages of puberty generally follow a set sequence:

- During the first stage of male puberty, the scrotum and testes grow larger.

- Next, the penis becomes longer, and the seminal vesicles and prostate gland grow.

- Hair begins to appear in the pubic area and later it grows on the face and underarms. During this time, a male's voice also deepens.

- Boys also undergo a growth spurt during puberty as they reach their adult height and weight.

Once a guy has reached puberty, he will produce millions of sperm cells every day. Each sperm is extremely small: only 1/600 of an inch (0.05 millimeters long). Sperm develop in the testicles within a system of tiny tubes called the seminiferous tubules. At birth, these tubules contain simple round cells, but during puberty, testosterone and other hormones cause these cells to transform into sperm cells. The cells divide and change until they have a head and short tail, like tadpoles. The head contains genetic material (genes). The sperm use their tails to push themselves into the epididymis, where they complete their development. It takes sperm about 4 to 6 weeks to travel through the epididymis.

The sperm then move to the vas deferens, or sperm duct. The seminal vesicles and prostate gland produce a whitish fluid called seminal fluid, which mixes with sperm to form semen when a male is sexually stimulated. The penis, which usually hangs limp, becomes hard when a male is sexually excited. Tissues in the penis fill with blood and it becomes stiff and erect (an erection). The rigidity of the erect penis makes it easier to insert into the female's vagina during sexual intercourse. When the erect penis is stimulated, muscles around the reproductive organs contract and force the semen through the duct system and urethra. Semen is pushed out of the male's body through his urethra — this process is called ejaculation. Each time a guy ejaculates, it can contain up to 500 million sperm.

When the male ejaculates during intercourse, semen is deposited into the female's vagina. From the vagina the sperm make their way up through the cervix and move through the uterus with help from uterine contractions. If a mature egg is in one of the female's fallopian tubes, a single sperm may penetrate it, and fertilization, or conception, occurs.

This fertilized egg is now called a zygote and contains 46 chromosomes — half from the egg and half from the sperm. The genetic material from the male and female has combined so that a new individual can be created. The zygote divides again and again as it grows in the female's uterus, maturing over the course of the pregnancy into an embryo, a fetus, and finally a newborn baby.

Problems Affecting the Male Reproductive System

Guys may sometimes experience reproductive system problems. Below are some examples of disorders that affect the male reproductive system:

Disorders of the Scrotum, Testicles or Epididymis

Conditions affecting the scrotal contents may involve the testicles, epididymis, or the scrotum itself.

- Testicular injury: Even a mild injury to the testicles can cause severe pain, bruising, or swelling. Most testicular injuries occur when the testicles are struck, hit, kicked, or crushed, usually during sports or due to other trauma. Testicular torsion, when one of the testicles twists around, cutting off its blood supply, can also happen to some guys. It's a serious problem that needs medical attention, but luckily it's not common.

- Varicocele: This is a varicose vein (an abnormally swollen vein) in the network of veins that run from the testicles. Varicoceles commonly develop while a guy is going through puberty. A varicocele is usually not harmful, although in some people it may damage the testicle or decrease sperm production, so it helps for a guy to see his doctor if he's concerned about changes in his testicles.

- Testicular cancer: This is one of the most common cancers in men younger than 40. It occurs when cells in the testicle divide abnormally and form a tumor. Testicular cancer can spread to other parts of the body, but if it's detected early, the cure rate is excellent. All guys should do testicular self-examinations regularly to help with early detection.

- Epididymitis is inflammation of the epididymis, the coiled tubes that connect the testes with the vas deferens. It is usually caused by an infection, such as the sexually transmitted disease chlamydia, and results in pain and swelling next to one of the testicles.

- Hydrocele: A hydrocele is when fluid collects in the membranes surrounding the testes. Hydroceles may cause swelling in the scrotum around the testicle but are generally painless. In some cases, surgery may be needed to correct the condition.

- Inguinal hernia: When a portion of the intestines pushes through an abnormal opening or weakening of the abdominal wall and into the groin or scrotum, it is known as an inguinal hernia. The hernia may look like a bulge or swelling in the groin area. It can be corrected with surgery.

Disorders of the Penis

Disorders affecting the penis include the following:

- Inflammation of the penis: Symptoms of penile inflammation include redness, itching, swelling, and pain. Balanitis is when the glans (the head of the penis) becomes inflamed. Posthitis is foreskin inflammation, which is usually due to a yeast or bacterial infection.

- Hypospadius is a disorder in which the urethra opens on the underside of the penis, not at the tip.

Female Reproductive System

The female reproductive system includes the ovaries, fallopian tubes, uterus, vagina, vulva, mammary glands and breasts. These organs are involved in the production and transportation of gametes and the production of sex hormones. The female reproductive system also facilitates the fertilization of ova by sperm and supports the development of offspring during pregnancy and infancy.

Anatomy of Female Reproductive System

Ovaries

The ovaries are a pair of small glands about the size and shape of almonds, located on the left and right sides of the pelvic body cavity lateral to the superior portion of the uterus. Ovaries produce female sex hormones such as estrogen and progesterone as well as ova (commonly called "eggs"), the female gametes. Ova are produced from oocyte cells that slowly develop throughout a woman's early life and reach maturity after puberty. Each month during ovulation, a mature ovum is released. The ovum travels from the ovary to the fallopian tube, where it may be fertilized before reaching the uterus.

Fallopian Tubes

The fallopian tubes are a pair of muscular tubes that extend from the left and right superior corners of the uterus to the edge of the ovaries. The fallopian tubes end in a funnel-shaped structure called the infundibulum, which is covered with small finger-like projections called fimbriae. The fimbriae swipe over the outside of the ovaries to pick up released ova and carry them into the infundibulum for transport to the uterus. The inside of each fallopian tube is covered in cilia that work with the smooth muscle of the tube to carry the ovum to the uterus.

Uterus

The uterus is a hollow, muscular, pear-shaped organ located posterior and superior to the urinary bladder. Connected to the two fallopian tubes on its superior end and to the vagina (via the cervix) on its inferior end, the uterus is also known as the womb, as it surrounds and supports the developing fetus during pregnancy. The inner lining of the uterus, known as the endometrium, provides support to the embryo during early development. The visceral muscles of the uterus contract during childbirth to push the fetus through the birth canal.

Vagina

The vagina is an elastic, muscular tube that connects the cervix of the uterus to the exterior of the body. It is located inferior to the uterus and posterior to the urinary bladder. The vagina functions as the receptacle for the penis during sexual intercourse and carries sperm to the uterus and fallopian tubes. It also serves as the birth canal by stretching to allow delivery of the fetus during childbirth. During menstruation, the menstrual flow exits the body via the vagina.

Vulva

The vulva is the collective name for the external female genitalia located in the pubic region of the body. The vulva surrounds the external ends of the urethral opening and the vagina and includes the mons pubis, labia majora, labia minora, and clitoris. The mons pubis, or pubic mound, is a raised layer of adipose tissue between the skin and the pubic bone that provides cushioning to the vulva. The inferior portion of the mons pubis splits into left and right halves called the labia majora. The mons pubis and labia majora are covered with pubic hairs. Inside of the labia majora are smaller, hairless folds of skin called the labia minora that surround the vaginal and urethral openings. On the superior end of the labia minora is a small mass of erectile tissue known as the clitoris that contains many nerve endings for sensing sexual pleasure.

Breasts and Mammary Glands

The breasts are specialized organs of the female body that contain mammary glands, milk ducts, and adipose tissue. The two breasts are located on the left and right sides of the thoracic region of the body. In the center of each breast is a highly pigmented nipple that releases milk when stimulated. The areola, a thickened, highly pigmented band of skin that surrounds the nipple, protects the underlying tissues during breastfeeding. The mammary glands are a special type of sudoriferous glands that have been modified to produce milk to feed infants. Within each breast, 15 to 20 clusters of mammary glands become active during pregnancy and remain active until milk is no longer needed. The milk passes through milk ducts on its way to the nipple, where it exits the body.

Physiology of Female Reproductive System

Reproductive Cycle

The female reproductive cycle is the process of producing an ovum and readying the uterus to receive a fertilized ovum to begin pregnancy. If an ovum is produced but not fertilized and implanted in the uterine wall, the reproductive cycle resets itself through menstruation. The entire reproductive cycle takes about 28 days on average, but may be as short as 24 days or as long as 36 days for some women.

Oogenesis and Ovulation

Under the influence of follicle stimulating hormone (FSH), and luteinizing hormone (LH), the ovaries produce a mature ovum in a process known as ovulation. By about 14 days into the reproductive cycle, an oocyte reaches maturity and is released as an ovum. Although the ovaries begin to mature many oocytes each month, usually only one ovum per cycle is released.

Fertilization

Once the mature ovum is released from the ovary, the fimbriae catch the egg and direct it down the fallopian tube to the uterus. It takes about a week for the ovum to travel to the uterus. If sperm are able to reach and penetrate the ovum, the ovum becomes a fertilized zygote containing a full complement of DNA. After a two-week period of rapid cell division known as the germinal period of development, the zygote forms an embryo. The embryo will then implant itself into the uterine wall and develop there during pregnancy.

Menstruation

While the ovum matures and travels through the fallopian tube, the endometrium grows and develops in preparation for the embryo. If the ovum is not fertilized in time or if it fails to implant into the endometrium, the arteries of the uterus constrict to cut off blood flow to the endometrium. The lack of blood flow causes cell death in the endometrium and the eventual shedding of tissue in a process known as menstruation. In a normal menstrual cycle, this shedding begins around day 28 and continues into the first few days of the new reproductive cycle.

Pregnancy

If the ovum is fertilized by a sperm cell, the fertilized embryo will implant itself into the endometrium and begin to form an amniotic cavity, umbilical cord, and placenta. For the first 8 weeks, the embryo will develop almost all of the tissues and organs present in the adult before entering the fetal period of development during weeks 9 through 38. During the fetal period, the fetus grows larger and more complex until it is ready to be born.

Lactation

Lactation is the production and release of milk to feed an infant. The production of milk begins prior to birth under the control of the hormone prolactin. Prolactin is produced in response to the suckling of an infant on the nipple, so milk is produced as long as active breastfeeding occurs. As soon as an infant is weaned, prolactin and milk production end soon after. The release of milk by the nipples is known as the "milk-letdown reflex" and is controlled by the hormone oxytocin. Oxytocin is also produced in response to infant suckling so that milk is only released when an infant is actively feeding.

Disorders of Female Reproductive System

Girls and women may sometimes experience reproductive system problems. Below are some examples of disorders that affect the female reproductive system:

With the Vulva and Vagina

- Vulvovaginitis, an inflammation of the vulva and vagina. It may be caused by irritating substances (such as laundry soaps or bubble baths). Poor personal hygiene (such as wiping from back to front after a bowel movement) may also cause this problem. Symptoms include redness and itching in the vaginal and vulvar areas and sometimes vaginal discharge. Vulvovaginitis can also be caused by an overgrowth of candida, a fungus normally present in the vagina.

- Nonmenstrual vaginal bleeding, most commonly due to the presence of a vaginal foreign body, often wadded-up toilet paper. It may also be due to urethral prolapse, a condition in which the mucous membranes of the urethra protrude into the vagina and form a tiny, donut-shaped mass of tissue that bleeds easily. It can also be due to a straddle injury (such as when falling onto a beam or bicycle frame) or vaginal trauma from sexual abuse.

With the Ovaries and Fallopian Tubes

- Ectopic pregnancy, when a fertilized egg, or zygote, doesn't travel into the uterus, but instead grows rapidly in the fallopian tube. Girls with this condition can develop severe abdominal pain and should see a doctor because surgery may be necessary.

- Endometriosis, when tissue normally found only in the uterus starts to grow outside the uterus — in the ovaries, fallopian tubes, or other parts of the pelvic cavity. It can cause abnormal bleeding, painful periods, and general pelvic pain.

- Ovarian tumors, although rare, can occur. Girls with ovarian tumors may have abdominal pain and masses that can be felt in the abdomen. Surgery may be needed to remove the tumor.

- Ovarian cysts, noncancerous sacs filled with fluid or semi-solid material. Although they are common and generally harmless, they can become a problem if they grow very large. Large cysts may push on surrounding organs, causing abdominal pain. In most cases, cysts will disappear on their own and treatment is unnecessary. If the cysts are painful, a doctor may prescribe birth control pills to alter their growth, or they may be removed by a surgeon.

- Polycystic ovary syndrome, a hormone disorder in which too many male hormones (androgens) are produced by the ovaries. This condition causes the ovaries to become enlarged and develop many fluid-filled sacs, or cysts. It often first appears during the teen years. Depending on the type and severity of the condition, it may be treated with drugs to regulate hormone balance and menstruation.

Menstrual Problems

A variety of menstrual problems can affect girls. Some of the more common conditions are:

- Dysmenorrhea, when a girl has painful periods.

- Menorrhagia, when a girl has a very heavy periods with excess bleeding.

- Oligomenorrhea, when a girl misses or has infrequent periods, even though she's been menstruating for a while and isn't pregnant.

- Amenorrhea, when a girl hasn't started her period by the time she is 16 years old or 3 years after starting puberty, has not developed signs of puberty by age 14, or has had normal periods but has stopped menstruating for some reason other than pregnancy.

Infections of the Female Reproductive System

- Sexually transmitted diseases (STDs): Also called sexually transmitted infections (STIs), these include pelvic inflammatory disease (PID), human immunodeficiency virus/acquired immunodeficiency syndrome (HIV/AIDS), human papillomavirus (HPV, or genital warts), syphilis, chlamydia, gonorrhea, and genital herpes (HSV). Most are spread from one person to another by sexual contact.

- Toxic shock syndrome: This uncommon but life-threatening illness is caused by toxins released into the body during a type of bacterial infection that is more likely to develop if a tampon is left in too long. It can produce high fever, diarrhea, vomiting, and shock.

11

Immune System

The immune system is a defense system that comprises of various biological structures and processes that enables an organism to fight against diseases and infections. This chapter elaborates the relevant aspects of the innate and adaptive immune system which will help in developing a holistic understanding of the human immune system.

The immune system (from the Latin word immunis, meaning: "free" or "untouched") protects the body like a guardian from harmful influences from the environment and is essential for survival. It is made up of different organs, cells and proteins and aside from the nervous system, it is the most complex system that the human body has.

As long as our body's system of defense is running smoothly, we do not notice the immune system. And yet, different groups of cells work together and form alliances against just about any pathogen (germ). But illness can occur if the performance of the immune system is compromised, if the pathogen is especially aggressive, or sometimes also if the body is confronted with a pathogen it has not come into contact before.

Tasks of the Immune System

Without an immune system, a human being would be just as exposed to the harmful influences of pathogens or other substances from the outside environment as to changes harmful to health happening inside of the body. The main tasks of the body's immune system are:

- Neutralizing pathogens like bacteria, viruses, parasites or fungithat have entered the body, and removing them from the body.

- Recognizing and neutralizing harmful substances from the environment.

- Fighting against the body's own cells that have changed due to an illness, for example cancerous cells.

The human immune system is divided into two broad groups called the Acquired Immune System and the Innate Immune System. These include:

- Lymphatics

- Lymph nodes

- Thymus

- Spleen

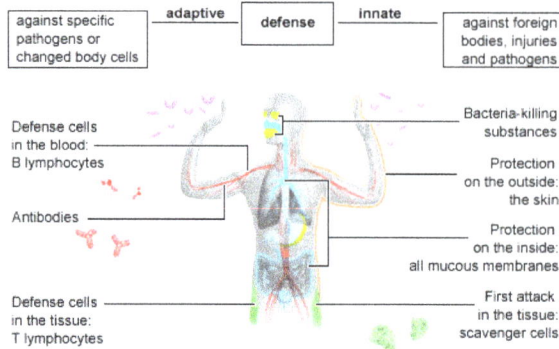

Different Types of Immunity

The immune system is divided into two parts, called the Acquired Immune System and the Innate Immune System. While each of these plays a role in defending the body, there are major differences between the two.

The innate immune system is always working to protect the body and does not require any special preparation to stop infection.

The acquired immune system needs to be 'primed' before it can work to its full effectiveness though, and is only really effective after it has seen a possible infective agent before.

An overview of these different systems is given in the chart below:

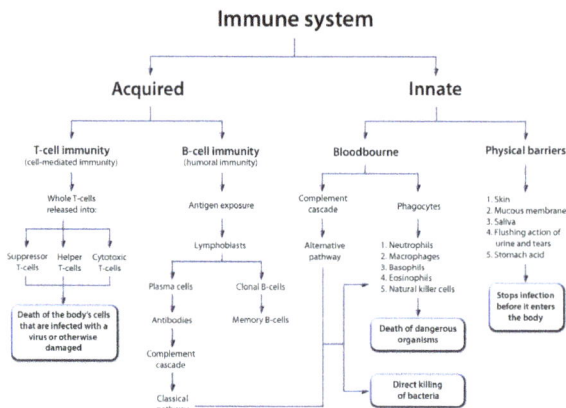

Structure and Organs of the Immune System

Lymphatic System

The lymphatic system is almost equivalent to the blood vessels, only instead of carrying blood through the body, the lymphatic system carries a substance called 'lymph'. Lymph is excess tissue fluid that has been drained from the body compartments. Lymphatic fluid is usually clear, watery, and has the same constitution as the blood, but without any cells. The lymphatic system is a complex network of lymphatic vessels (that carry the lymph), along which there are occasional lymph nodes. After the lymphatic system has collected all the lymph, this passes through the lymph nodes before being put back into the blood via a large vein just below the neck. In the lymphatic system there are lots of cells called lymphocytes (the T and B cells) that circulate around and are part of the acquired immune system.

Lymphoid Tissue

Lymphoid tissue is scattered throughout the body and is home to the lymphocytes. Lymphocytes are packed into clusters in the walls of parts of the body that are often exposed to foreign substances. These sites include the gastrointestinal system as well as the tonsils which play a role in protecting the body from any air-borne infections.

Lymph Nodes

Lymph nodes are small, oval structures that can be anywhere from 1mm to 25mm big. Blood vessels and nerves attach to the lymph nodes, as well as two sets of lymphatic vessels – those that enter the lymph node and those that leave it.

The lymph enters from one side and slowly moves past all the cells of the lymph node before leaving through the other lymphatic vessel. This allows the lymph time to access as many of the lymphatic cells as possible. In the lymph node there is a dense packaging of immune cells such as macrophages. These are the 'big eaters' and will engulf and destroy anything dangerous that they can. They also play a role in showing these substances to the T and B cells. There are also areas of the lymph node called 'germinal centres' where all the b cells multiply to fight off infection. In another part of the lymph node, there are mostly T cells. When they need to, the lymphocytes leave the lymph node and enter the circulation to fight infection.

The lymph nodes are there as a filter for the lymph before it re-enters the venous system. 99% of all the foreign substances that arrive at the lymph node are removed. Lymph nodes are found in regions such as the base of the neck, the armpit and the groin. Swelling or inflammation of these nodes is usually in response to an infection in one of the areas that is drained by the lymph node. This is often what is meant when someone says that they have 'swollen glands'.

Thymus

The thymus is a lymphoid organ located in the lower part of the neck and the front of the chest. With age, the thymus becomes smaller and loses most of its active immune cells. The outside of the thymus contains lymphoid stem cells (which are immature cells, still capable of growing) that divide rapidly, producing cells that mature into T cells. These T cells then migrate to the middle of the thymus. There are also cells in the thymus that release hormones (signalling chemicals) that cause T cells to grow.

Spleen

The spleen is the largest of the lymphoid organs. It is usually purple in colour, and located in the upper-left of the abdomen (the belly). The spleen is located behind the stomach, in front of the diaphragm (the muscle used for breathing), and next to the left kidney. The spleen can vary in size and shape dramatically; however, it is usually about 12cm long and 7cm wide (about the size of a clenched fist). The spleen contains large amounts of blood that is periodically pushes into the circulation by contraction of some tiny muscles that surround it.

There are two different 'parts' to the spleen, each with a different function. The 'red pulp' is named because of its color and its role is to filter the blood. It does this by having tiny holes in its blood vessels that only allow some types of blood cells through. The blood cells that are a little older or in any way defective are not flexible enough to squeeze through these holes and so gets stuck. These stuck cells are then eaten by the macrophages.

The 'white pulp' is basically areas of lymphoid tissue in the middle of the spleen. There are areas filled with T cells and B cells. These make up about 5-20% of the spleen. There are lots more of the B cells in younger people than there are in older people, and their numbers in the spleen decrease with age.

Innate Immune System

The innate immune system is the phylogenically oldest component of the human immune system. The innate immune system is made of defenses against infection that can be activated immediately once a pathogen attacks. The innate immune system is essentially made up of barriers that aim to keep viruses, bacteria, parasites, and other foreign particles out of your body or limit their ability to spread and move throughout the body. The innate immune system includes:

- Physical Barriers - such as skin, the gastrointestinal tract, the respiratory tract, the nasopharynx, cilia, eyelashes and other body hair.

- Defense Mechanisms - such as secretions, mucous, bile, gastric acid, saliva, tears, and sweat.

- General Immune Responses - such as inflammation, complement, and non-specific cellular responses. The inflammatory response actively brings immune cells to the site of an infection by increasing blood flow to the area. Complement is an immune response that marks pathogens for destruction and makes holes in the cell membrane of the pathogen.

The innate immune system is always general, or *nonspecific*, meaning anything that is identified as foreign or *non-self* is a target for the innate immune response. The innate immune system is activated by the presence of antigens and their chemical properties.

Cells of the Innate Immune System

There are many types of white blood cells, or *leukocytes*, that work to defend and protect the human body. In order to patrol the entire body, leukocytes travel by way of the circulatory system.

The following cells are leukocytes of the innate immune system:

- Phagocytes, or Phagocytic cells: Phagocyte means "eating cell", which describes what role phagocytes play in the immune response. Phagocytes circulate throughout the body, looking for potential threats, like bacteria and viruses, to engulf and destroy.

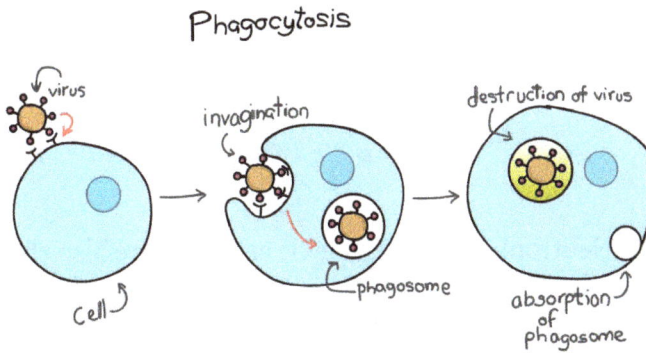

Phagocytosis

This section explains how phagocytes know what to engulf, and how phagocytosis works.

- Macrophages: Macrophages, commonly abbreviated as "Mφ", are efficient phagocytic cells that can leave the circulatory system by moving across the walls of capillary vessels. The ability to roam outside of the circulatory system is important, because it allows macrophages to hunt pathogens with less limits. Macrophages can also release cytokines in order to signal and recruit other cells to an area with pathogens.

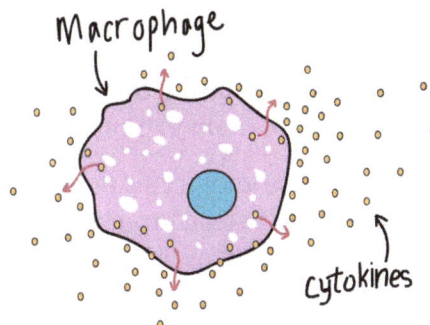

Macrophage

- Mast cells: Mast cells are found in mucous membranes and connective tissues, and are important for wound healing and defense against pathogens via the inflammatory response. When mast cells are activated, they release cytokines and granules that contain chemical molecules to create an inflammatory cascade. Mediators, such as histamine, cause blood vessels to dilate, increasing blood flow and cell trafficking to the area of infection. The cytokines released during this process act as a messenger service, alerting other immune cells, like neutrophils and macrophages, to make their way to the area of infection, or to be on alert for circulating threats.

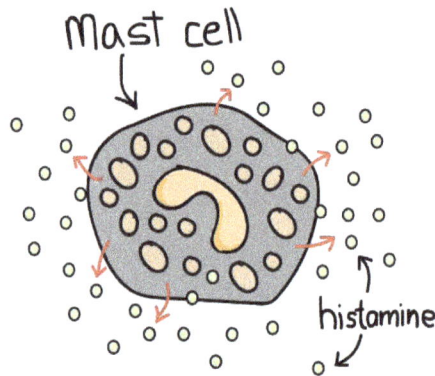

Mast cell

histamine

- Neutrophils: Neutrophils are phagocytic cells that are also classified as *granulocytes* because they contain granules in their cytoplasm. These granules are very toxic to bacteria and fungi, and cause them to stop proliferating or die on contact.

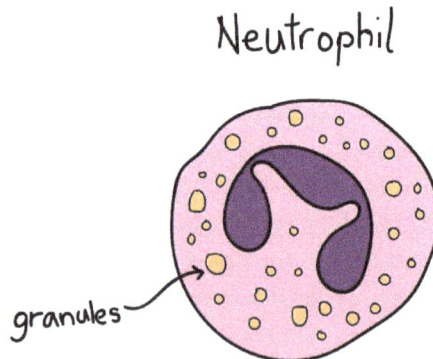

Neutrophil

granules

The bone marrow of an average healthy adult makes approximately 100 billion new neutrophils per day. Neutrophils are typically the first cells to arrive at the site of an infection because there are so many of them in circulation at any given time.

- Eosinophils: Eosinophils are granulocytes target multicellular parasites. Eosinophils secrete a range of highly toxic proteins and free radicals that kill bacteria and parasites. The use of toxic proteins and free radicals also causes tissue damage during allergic reactions, so activation and toxin release by eosinophils is highly regulated to prevent any unnecessary tissue damage.

While eosinophils only make up 1-6% of the white blood cells, they are found in many locations, including the thymus, lower gastrointestinal tract, ovaries, uterus, spleen, and lymph nodes.

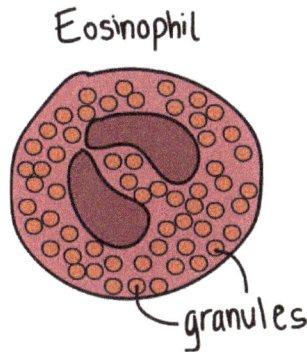

Eosinophil

granules

- Basophils: Basophils are also granulocytes that attack multicellular parasites. Basophils release histamine, much like mast cells. The use of histamine makes basophils and mast cells key players in mounting an allergic response.

- Natural killer cells: Natural killer cells (NK cells), do not attack pathogens directly. Instead, natural killer cells destroy infected host cells in order to stop the spread of an infection. Infected or compromised host cells can signal natural kill cells for destruction through the expression of specific receptors and antigen presentation.

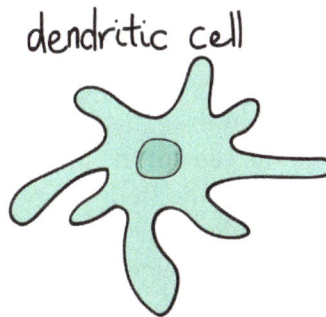

dendritic cell

- Dendritic cells: Dendritic cells are antigen-presenting cells that are located in tissues, and can contact external environments through the skin, the inner mucosal lining of the nose, lungs, stomach, and intestines. Since dendritic cells are located in tissues that are common points for initial infection, they can identify threats and act as messengers for the rest of the immune system by antigen presentation. Dendritic cells also act as bridge between the innate immune system and the adaptive immune system.

Complement System

The complement system (also called the complement cascade) is a mechanism that complements other aspects of the immune response. Typically, the complement system acts as a part of the innate immune system, but it can work with the adaptive immune system if necessary.

The complement system is made of a variety of proteins that, when inactive, circulate in the blood. When activated, these proteins come together to initiate the complement cascade, which starts the following steps:

1. Opsonization: Opsonization is a process in which foreign particles are marked for phagocytosis. All of the pathways require an antigen to signal that there is a threat present. Opsonization tags infected cells and identifies circulating pathogens expressing the same antigens.

2. Chemotaxis: Chemotaxis is the attraction and movement of macrophages to a chemical signal. Chemotaxis uses cytokines and chemokines to attract macrophages and neutrophils to the site of infection, ensuring that pathogens in the area will be destroyed. By bringing immune cells to an area with identified pathogens, it improves the likelihood that the threats will be destroyed and the infection will be treated.

3. Cell Lysis: Lysis is the breaking down or destruction of the membrane of a cell. The proteins of the complement system puncture the membranes of foreign cells, destroying the integrity of the pathogen. Destroying the membrane of foreign cells or pathogens weakens their ability to proliferate, and helps to stop the spread of infection.

4. Agglutination: Agglutination uses antibodies to cluster and bind pathogens together, much like a cowboy rounds up his cattle. By bringing as many pathogens together in the same area, the cells of the immune system can mount an attack and weaken the infection. Other innate immune system cells continue to circulate throughout the body in order to track down any other pathogens that have not been clustered and bound for destruction.

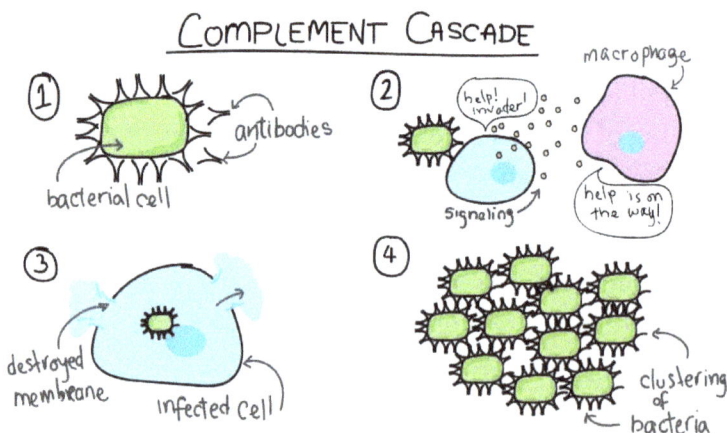

COMPLEMENT CASCADE

Complement Cascade

The steps of the complement cascade facilitate the search for and removal of antigens by placing them in large clumps, making it easier for other aspects of the immune system to do their jobs. Remember that the complement system is a supplemental cascade of proteins that assists, or "complements" the other aspects of the innate immune system.

The innate immune system works to fight off pathogens before they can start an active infection. For some cases, the innate immune response is not enough, or the pathogen is able to exploit the innate immune response for a way into the host cells. In such situations, the innate immune system works with the adaptive immune system to reduce the severity of infection, and to fight off any additional invaders while the adaptive immune system is busy destroying the initial infection.

Adaptive Immune System

The adaptive immune system, also called acquired immunity, uses specific antigens to strategically mount an immune response. Unlike the innate immune system, which attacks only based on the identification of general threats, the adaptive immunity is activated by exposure to pathogens, and uses an immunological memory to learn about the threat and enhance the immune response accordingly. The adaptive immune response is much slower to respond to threats and infections than the innate immune response, which is primed and ready to fight at all times.

Cells of the Adaptive Immune System

Unlike the innate immune system, the adaptive immune system relies on fewer types of cells to carry out its tasks: *B cells* and *T cells*.

Both B cells and T cells are lymphocytes that are derived from specific types of stem cells, called multipotent hematopoietic stem cells, in the bone marrow. After they are made in the bone marrow, they need to mature and become activated. Each type of cell follows different paths to their final, mature forms.

B cells

After formation and maturation in the bone marrow (hence the name "B cell"), the naive *B cells* move into the lymphatic system to circulate throughout the body. In the lymphatic system, naive B cells encounter an antigen, which starts the maturation process for the B cell. B cells each have one of millions of distinctive surface antigen-specific receptors that are inherent to the organism's DNA. For example, naive B cells express antibodies on their cell surface, which can also be called membrane-bound antibodies.

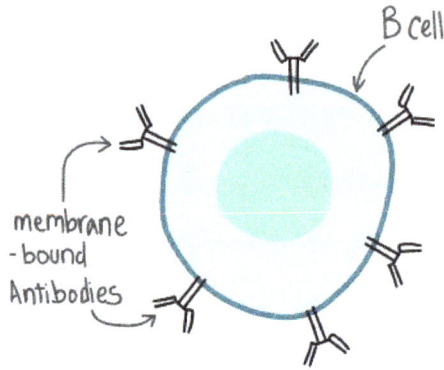

When a naive B cell encounters an antigen that fits or matches its membrane-bound antibody, it quickly divides in order to become either a *memory B cell* or an *effector B cell*, which is also called a *plasma cell*. Antibodies can bind to antigens directly.

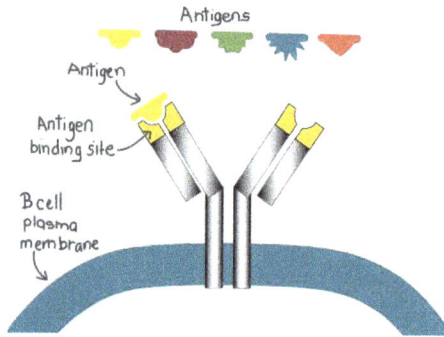

The antigen must effectively bind with a naive B cell's membrane-bound antibody in order to set off differentiation, or the process of becoming one of the new forms of a B cell.

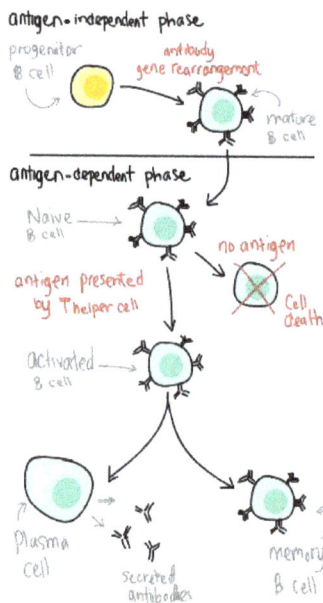

Memory B cells express the same membrane-bound antibody as the original naive B cell, or the "parent B cell". Plasma B cells produce the same antibody as the parent B cell, but they aren't membrane bound. Instead, plasma B cells can secrete antibodies. Secreted antibodies work to identify free pathogens that are circulating throughout the body. When the naive B cell divides and differentiates, both plasma cells and memory B cells are made.

B cells also express a specialized receptor, called the *B cell receptor (BCR)*. B cell receptors assist with antigen binding, as well as internalization and processing of the antigen. B cell receptors also play an important role in signaling pathways. After the antigen is internalized and processed, the B cell can initiate signaling pathways, such as cytokine release, to communicate with other cells of the immune system.

T cells

Once formed in the bone marrow, *T progenitor cells* migrate to the thymus (hence the name "T cell") to mature and become T cells. While in the thymus, the developing T cells start to express T cell receptors *(TCRs)* and other receptors called *CD4* and *CD8* receptors. All T cells express T cell receptors, and either CD4 or CD8, not both. So, some T cells will express CD4, and others will express CD8.

Unlike antibodies, which can bind to antigens directly, T cell receptors can only recognize antigens that are bound to certain receptor molecules, called Major Histocompatibility Complex class 1 (MHCI) and class 2 (MHCII). These MHC molecules are membrane-bound surface receptors on *antigen-presenting cells*, like dendritic cells and macrophages. CD4 and CD8 play a role in T cell recognition and activation by binding to either MHCI or MHCII.

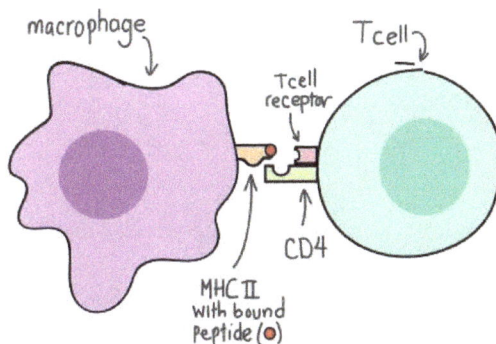

T cell receptors have to undergo a process called rearrangement, causing the nearly limitless recombination of a gene that expresses T cell receptors. The process of rearrangement allows for a lot of binding diversity. This diversity could potentially lead to accidental attacks against self cells and molecules because some rearrangement configurations can accidentally mimic a person's self molecules and proteins. Mature T cells should recognize only foreign antigens combined with self-MHC molecules in order to mount an appropriate immune response.

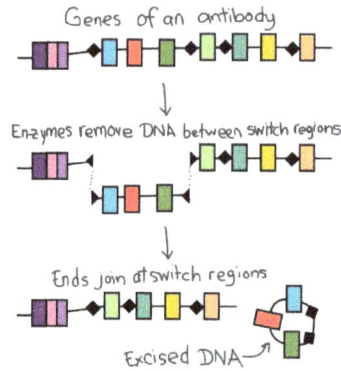

Genes of an antibody

Enzymes remove DNA between switch regions

Ends join at switch regions

Excised DNA

In order to make sure T cells will perform properly once they have matured and have been released from the thymus, they undergo two selection processes:

1. Positive selection ensures MHC restriction by testing the ability of MHCI and MHCII to distinguish between self and nonself proteins. In order to pass the positive selection process, cells must be capable of binding only self-MHC molecules. If these cells bind nonself molecules instead of self-MHC molecules, they fail the positive selection process and are eliminated by apoptosis.

2. Negative selection tests for self tolerance. Negative selection tests the binding capabilities of CD4 and CD8 specifically. The ideal example of self tolerance is when a T cell will only bind to self-MHC molecules presenting a foreign antigen. If a T cell binds, via CD4 or CD8, a self-MHC molecule that isn't presenting an antigen, or a self-MHC molecule that presenting a self-antigen, it will fail negative selection and be eliminated by apoptosis.

These two selection processes are put into place to protect your own cells and tissues against your own immune response. Without these selection processes, autoimmune diseases would be much more common.

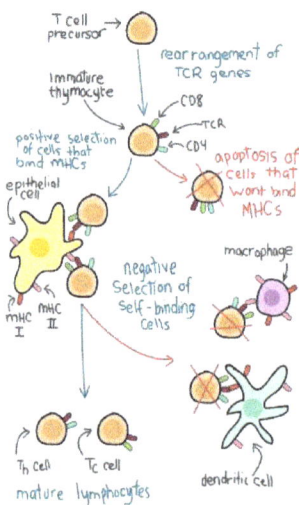

After positive and negative selection, we are left with three types of mature T cells: Helper T cells (T_H start subscript, H, end subscript cells), Cytotoxic T cells (T_C start subscript, C, end subscript cells), and T regulatory cells (T_{reg} start subscript, r, e, g, end subscriptcells).

- Helper T cells express CD4, and help with the activation of T_C start subscript, C, end subscript cells, B cells, and other immune cells.

- Cytotoxic T cells express CD8, and are responsible for removing pathogens and infected host cells.

- T regulatory cells express CD4 and another receptor, called CD25. T regulatory cells help distinguish between self and nonself molecules, and by doing so, reduce the risk of autoimmune diseases.

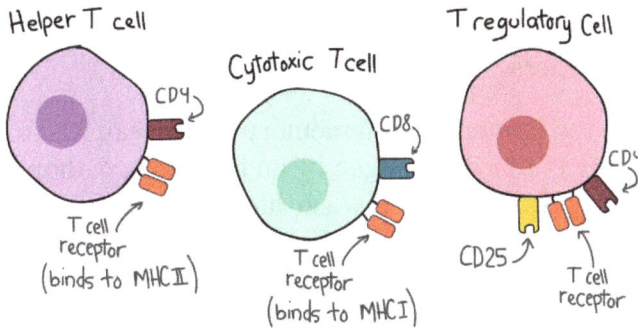

Humoral vs. Cell Mediated Immunity

Immunity refers to the ability of your immune system to defend against infection and disease. There are two types of immunity that the adaptive immune system provides, and they are dependent on the functions of B and T cells.

Humoral immunity is immunity from serum antibodies produced by plasma cells. More specifically, someone who has never been exposed to a specific disease can gain humoral immunity through administration of antibodies from someone who has been exposed, and survived the same disease. "Humoral" refers to the bodily fluids where these free-floating serum antibodies bind to antigens and assist with elimination.

Cell-mediated immunity can be acquired through T cells from someone who is immune to the target disease or infection. "Cell-mediated" refers to the fact that the response is carried out by cytotoxic cells. Much like humoral immunity, someone who has not been exposed to a specific disease can gain cell-mediated immunity through the administration of T_H and T_C cells from someone that has been exposed, and survived the same disease. The T_H cells act to activate other immune cells, while the T_C cells assist with the elimination of pathogens and infected host cells.

Immunological Memory

Because the adaptive immune system can learn and remember specific pathogens, it can provide long-lasting defense and protection against recurrent infections. When the adaptive immune system is exposed to a new threat, the specifics of the antigen are memorized so we are prevented from getting the disease again. The concept of immune memory is due to the body's ability to make antibodies against different pathogens.

A good example of immunological memory is shown in vaccinations. A vaccination against a virus can be made using either active, but weakened or attenuated virus, or using specific parts of the virus that are not active. Both attenuated whole virus and virus particles cannot actually cause an active infection. Instead, they mimic the presence of an active virus in order to cause an immune response, even though there are no real threats present. By getting a vaccination, you are exposing your body to the antigen required to produce antibodies specific to that virus, and acquire a memory of the virus, without experiencing illness.

Some breakdowns in the immunological memory system can lead to autoimmune diseases. Molecular mimicry of a self-antigen by an infectious pathogen, such as bacteria and viruses, may trigger autoimmune disease due to a cross-reactive immune response against the infection. One example of an organism that uses molecular mimicry to hide from immunological defenses is Streptococcus infection.

Permissions

All chapters in this book are published with permission under the Creative Commons Attribution Share Alike License or equivalent. Every chapter published in this book has been scrutinized by our experts. Their significance has been extensively debated. The topics covered herein carry significant information for a comprehensive understanding. They may even be implemented as practical applications or may be referred to as a beginning point for further studies.

We would like to thank the editorial team for lending their expertise to make the book truly unique. They have played a crucial role in the development of this book. Without their invaluable contributions this book wouldn't have been possible. They have made vital efforts to compile up to date information on the varied aspects of this subject to make this book a valuable addition to the collection of many professionals and students.

This book was conceptualized with the vision of imparting up-to-date and integrated information in this field. To ensure the same, a matchless editorial board was set up. Every individual on the board went through rigorous rounds of assessment to prove their worth. After which they invested a large part of their time researching and compiling the most relevant data for our readers.

The editorial board has been involved in producing this book since its inception. They have spent rigorous hours researching and exploring the diverse topics which have resulted in the successful publishing of this book. They have passed on their knowledge of decades through this book. To expedite this challenging task, the publisher supported the team at every step. A small team of assistant editors was also appointed to further simplify the editing procedure and attain best results for the readers.

Apart from the editorial board, the designing team has also invested a significant amount of their time in understanding the subject and creating the most relevant covers. They scrutinized every image to scout for the most suitable representation of the subject and create an appropriate cover for the book.

The publishing team has been an ardent support to the editorial, designing and production team. Their endless efforts to recruit the best for this project, has resulted in the accomplishment of this book. They are a veteran in the field of academics and their pool of knowledge is as vast as their experience in printing. Their expertise and guidance has proved useful at every step. Their uncompromising quality standards have made this book an exceptional effort. Their encouragement from time to time has been an inspiration for everyone.

The publisher and the editorial board hope that this book will prove to be a valuable piece of knowledge for students, practitioners and scholars across the globe.

Index